Energy-Efficient Industrial Systems

Energy-Efficient Industrial Systems

Evaluation and Implementation

Dr. Lal Jayamaha

New York   Chicago   San Francisco   Athens
London   Madrid   Mexico City   Milan
New Delhi   Singapore   Sydney   Toronto

Library of Congress Control Number: 2016934577

Energy-Efficient Industrial Systems: Evaluation and Implementation

1 2 3 4 5 6 7 8 9 0   DOC/DOC   1 2 1 0 9 8 7 6

ISBN 978-1-25-958978-2
MHID 1-25-958978-1

This book is printed on acid-free paper.

Sponsoring Editor	**Acquisitions Coordinator**	**Proofreader**
Robert Argentieri	Lauren Rogers	Connie Blazewicz
Editing Supervisor	**Project Manager**	**Art Director, Cover**
Stephen M. Smith	Tanya Punj, Cenveo® Publisher Services	Jeff Weeks
Production Supervisor		**Composition**
Lynn M. Messina	**Copy Editor**	Cenveo Publisher Services
	Anirban Mittra, Cenveo Publisher Services	

This book is dedicated to the memory
of my late parents, George and Marie Jayamaha.
Thank you for your love and kindness
and for knowing the importance
of a good education.

About the Author

Dr. Lal Jayamaha is the founder of LJ Energy, a leading energy services company in Asia. Over the last 20 years, he has been the team leader for more than 100 energy efficiency projects involving commercial buildings and industrial facilities. In addition, he has successfully completed many green building projects. Dr. Jayamaha earned a Ph.D. degree from the National University of Singapore and is a Chartered Engineer (UK), a Professional Engineer (Singapore), and a LEED AP. He is also a Certified Energy Manager, a member of ASHRAE, and a member of the Institution of Engineers (UK).

Contents

Preface

Industrial plants are significant users of energy, which is normally supplied in the form of electricity or fuel. Energy is used for operating systems such as boilers, air compressors, pumps, fans, and refrigeration equipment in industrial facilities and often account for about half their operating cost. Based on my past experience, having completed many industrial energy efficiency projects, there is often potential to reduce the overall energy consumption of industrial facilities by 10% to 20% by optimizing various systems.

This book contains 10 self-contained chapters on the different aspects and systems relating to energy consumption in industrial facilities. The book is written as a "practical" guide for readers interested in energy savings and contains many actual examples from my past work. Further, the energy-saving strategies and measures described in the book have been implemented before and therefore are of a practical nature and can be implemented by readers based on their individual needs.

This book on the energy efficiency of industrial systems is written mainly for energy managers and professionals who are involved in operations, maintenance, and energy management in industrial plants, and professionals working in the energy efficiency industry who would like to broaden their knowledge about the subject.

Chapter 1 provides an introduction to the subject of energy management and outlines the important aspects of an energy management program. Thereafter, the subsequent chapters describe the different systems and aspects of energy consumption in industrial facilities. Each chapter includes a brief introduction to the subject to help readers who may not be familiar with some of the basic concepts associated with each topic.

Chapter 2 on motors and drives describes the various types of motors commonly used and the various opportunities for improving the operating efficiency of motor-driven systems that account for the majority of electrical energy consumption in industrial plants. Chapters 3 and 4 provide an overview of various pumping and fan systems and how to optimize them. Some of the topics covered are system characteristics, system losses, affinity laws, and design best practices.

Chapter 5 is on boilers and steam systems. The chapter includes an introduction to the properties of steam followed by some of the important aspects such as boiler efficiency. Thereafter, various potential energy-saving measures that can improve the efficiency of steam systems through operational strategies and system design are described. Most industrial facilities require cooling systems for process cooling and for storage of raw materials or finished products. Many cooling systems operate at relatively low temperatures and therefore consume a significant amount of energy. Chapter 6 includes measures on how to improve the energy efficiency of industrial cooling systems such as cooling towers, chillers, and refrigeration plants.

Compressed air systems are one of the most energy-intensive systems and are covered in Chap. 7. A brief introduction to compressed air systems is followed by a number of commonly applicable energy-saving measures that can lead to 20% to 30% reduction in energy usage of such systems.

Often, industrial processes require heating as well as cooling, both of which require energy for their operation. Since cooling requires removing heat while heating involves adding heat, there are many opportunities to reuse the heat removed from one process to heat another process, which helps to minimize energy usage. Various heat recovery systems that can be adopted for such applications are presented in Chap. 8.

Many industrial facilities require both electricity and heating, which can be generated simultaneously using combined heat and power systems, which are relatively much more efficient than the conventional method of purchasing electricity from the grid and supplying heating requirements using boilers or furnaces. Chapter 9 explains various combined heat and power systems with examples on how to evaluate the feasibility of implementing such a system. Finally, Chap. 10 outlines many of the criteria used to evaluate the financial viability of energy management projects. In addition to the commonly used criteria such as simple payback period and return on investment, other concepts such as net present value, internal rate of return, sensitivity analysis, and life cycle cost are discussed in the chapter.

I would like to take this opportunity to thank my present and past colleagues, without whom I would not have been able to successfully complete so many energy audits and energy efficiency projects. Also, special thanks to Dr. Raisul for preparing some of the diagrams used in the book.

Dr. Lal Jayamaha

Abbreviations and Acronyms

AC	alternating current
ACFM	actual cubic feet per minute
ASHRAE	American Society of Heating, Refrigerating, and Air-Conditioning Engineers
ASME	American Society of Mechanical Engineers
CAGI	Compressed Air and Gas Institute
CEMP	European Committee of Manufacturers of Electrical Machines and Power Electronics
CHP	combined heat and power
CHPC	combined heat power and cooling
CO_2	carbon dioxide
COP	coefficient of performance
DC	direct current
DX	direct expansion
EAC	energy accounting center
EEI	energy efficiency index
EnPI	energy performance indicator
EPAct	Energy Policy Act of 1992
FAD	free air delivery
FRP	fiber-reinforced plastic
FV	future value
HRSG	heat recovery steam generator
Hz	hertz
IEC	International Electrotechnical Commission
IGA	investment-grade audit
IRR	internal rate of return
kW	kilowatt
kWh	kilowatt hour
LCC	life cycle cost
LMTD	log mean temperature difference
NEMA	National Electrical Manufacturers Association

NPV	net present value
O_2	oxygen
PAM	pulse amplitude modulation
PV	present value
PWM	pulse width modulation
RH	relative humidity
ROI	return on investment
rpm	revolutions per minute
RT	refrigeration tons
RTD	resistance temperature detector
SCFM	standard cubic feet per minute
SEC	specific energy consumption
TDS	total dissolved solids
VFD	variable-frequency drive
VSD	variable-speed drive

Energy-Efficient Industrial Systems

CHAPTER 1

Energy Management

1.1 Introduction

Energy management is a process used by many organizations, primarily, to reduce energy consumption and energy cost. The approach is similar to other good management practices like quality management, environmental management, and health and safety management adopted by organizations to improve and maintain their various management processes.

Energy management is a procedure to manage the energy consumption in an organization by ensuring that energy is used efficiently and without wastage. It is a systematic approach to ensure that energy management is sustainable.

Figure 1.1 shows a typical energy consumption trend for an organization which does not have a sustainable energy management program.

In such an organization, the energy consumption will tend to increase over time and when it reaches a value high enough to capture the attention of the management team, action will be taken to reduce the usage of energy. However, over time, when energy consumption has reduced sufficiently, less priority will be given to energy management and the energy usage will rise gradually. This cycle will continue with ups and downs in energy usage, unless a sustainable energy management program is implemented.

The following example illustrates what could happen in the absence of a good energy management system. Say an engineer optimizes a particular pumping system and sets the pump capacity using a variable-frequency drive (VFD) to achieve the required liquid flow rate. Then, a few years later, a new engineer increases the pump VFD set-point in response to a complaint of insufficient flow rate from a particular user. However, if there were a formal energy management system with documentation, the engineer would first investigate the problem and probably solve it without compromising on the energy consumption.

Figure 1.2 shows a common energy consumption profile for an organization which has implemented an effective energy management program where after the initial reduction in energy use, the usage rate is maintained and reduced further over time.

Once the energy consumption level is reduced in an organization with a good energy management program, the best practices incorporated in the program are able to maintain this level. Further reduction in energy usage is possible by other strategies such as the adoption of new technologies.

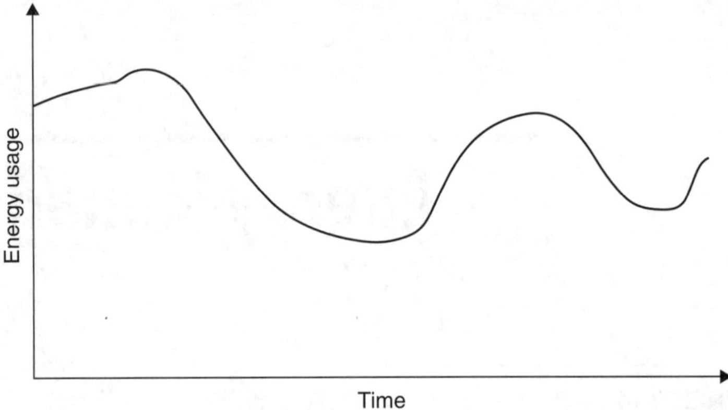

Figure 1.1   Typical energy consumption profile for an organization without an energy management program.

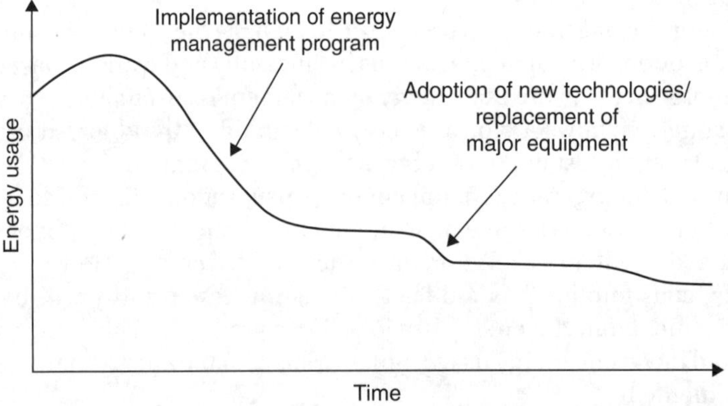

Figure 1.2   Possible energy consumption profile for an organization with a sustainable energy management program.

1.2   Main Drivers for Energy Management

Although the main motivation for energy management can differ from one organization to another, the main drivers can generally be classified as follows:

- Financial
- Environmental
- Organizational

The most common driver is financial, which relates to reducing the operating cost of an organization. Due to economic reasons, organizations around the world are constantly under pressure to reduce energy cost, which is one of the main cost drivers for businesses. The reduction in energy consumption leads to reduction in operating costs and thereby helps to improve the profitability of organizations.

Although in the past, the leading and probably only driver for energy management was financial, in the last decade, concerns relating to the environment have become more dominant. The main environmental concern relating to energy consumption is the emission of carbon dioxide (CO_2), which is a "greenhouse gas" that contributes to global warming. Due to the release of CO_2 during the burning of fossil fuels, CO_2 emissions can be closely correlated to energy use.

Another environment-related concern is the ever-increasing demand for fossil fuels, such as oil and gas, to support economic development. Since these are nonrenewable energy sources that take millions of years to form, they draw on finite resources, which will eventually be depleted.

Although financial and environmental aspects are the most common drivers for most, there is a trend for some to implement energy management programs purely for what can be called organizational drivers. These organization-related reasons can be corporate social responsibility (CSR) policies or statutory requirements imposed by the various jurisdictions. Often, energy management programs implemented purely to satisfy such requirements end up being merely for compliance, leading to little or no real reduction in energy consumption. Another reason for implementing an energy management program which can be classified as an organizational driver is when specific requirements are imposed by customers on their suppliers such as the need for them to be certified to an international energy management standard or having to provide periodic reporting of energy use and carbon emissions.

1.3   Typical Energy Management Program

An energy management program is able to achieve its objectives by means of the following key actions:

- Securing top management support for the program
- Assigning responsibilities to key personnel
- Creating awareness and educating all stakeholders on reducing energy usage
- Establishing targets for energy usage
- Developing plans for reducing energy usage
- Continuous monitoring of energy usage
- Reporting of energy usage
- Continuous reviewing of and implementing improvements to established plans

A typical energy management program can be expressed in terms of specific tasks or generalized processes, as depicted in Figs. 1.3 and 1.4.

The traditional approach is shown in Fig. 1.3 in which the first step is to select a team with the right skills to lead and execute the energy management program. Thereafter, the energy management team needs to set the objectives and priorities for the

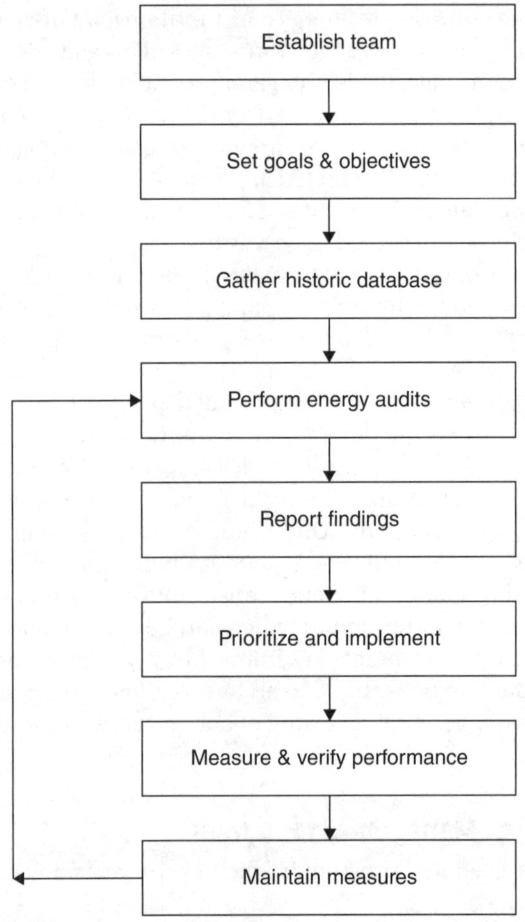

FIGURE 1.3 Traditional approach to energy management.

energy management program. Some of the aspects that would be covered are targeted savings, available budgets, and time frame for implementation.

Historic data such as utility consumption, occupancy rates, operating hours, occupied building floor areas, and production volumes need to be compiled so that future energy savings achieved can be evaluated by comparing with this database.

The most important part of the energy management program is to identify specific energy-saving measures. This is achieved through an energy audit. Once the audit is completed, the reported findings are used to decide on the energy-saving measures to be implemented (prioritizing where necessary).

Once the identified measures are implemented, the postinstallation performance is monitored to ensure that projected savings are achieved. In the event that savings fall short of the targeted values, further investigation is carried out to find the reasons for the shortfall and to identify possible remedial measures.

Finally, once all the measures are implemented, they need to be maintained to ensure that the achieved savings can be sustained over the expected life of the system.

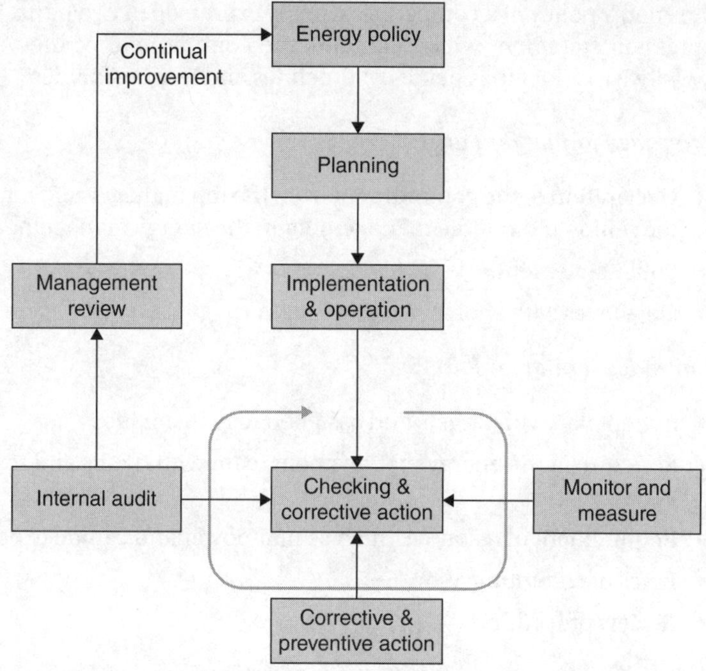

FIGURE 1.4   Plan–Do–Check–Act approach.

The newer approach adopted in ISO 50001 (International Standard for Energy Management System) follows the Plan–Do–Check–Act (PDCA) framework. The initial step of "Plan" is to establish the objectives and goals for the improvement in energy performance and company policy to achieve them.

The "Do" involves implementing the established plan, while the "Check" process involves monitoring the actual performance against policies, objectives, and key performance indicators, and reporting the results.

The final step, "Act," is to take necessary action to continually improve the energy performance of the organization.

1.4   Energy Policy

Energy policy of an organization should be formulated with an aim to fulfill the requirements necessary for achieving the targeted energy performance, keeping in mind the nature and scale of the organization's energy use. It should also outline the scope and boundaries of the energy management system.

The key contents of an energy policy comprise the objectives and targets. Energy policy should also state the commitment to continual improvement in energy performance while complying with all applicable legal and other requirements.

The energy policy should be documented, communicated, and understood by all the key stakeholders and be regularly reviewed and improved.

The energy policy of a company is usually expressed in two parts. Part 1 consists of an expression statement which contains the commitment of the management and goals, while Part 2 includes details of timelines and responsibilities.

Recommended format for Part 1:

- Declaration of the commitment from the top management and involvement of the senior and middle management in the energy management program
- Policy statement
- Objectives with short-term and long-term goals

Recommended format for Part 2:

- Action plan with targets and assigned responsibilities
- Structure of the energy management team with duties and responsibilities for each member
- Requirement of resources such as manpower and capital expenditure
- Lines of communication
- Review procedures

An example of a company energy policy (Part 1):

ABC company aims to reduce the environmental impact by implementing an energy management program compliant with ISO 50001. The company will continually strive to reduce the annual energy consumption by x%. All manufacturing processes in the organization will be designed, maintained, and operated to optimize energy usage. This policy shall be communicated to all employees and stakeholders. This policy will be reviewed annually.

(Name & signature of company chief executive)

1.5   Energy Management Team

Since energy management requires the commitment from the entire organization, a committee or team which can represent everyone is required to successfully implement an energy management program.

Team members are usually chosen from different departments within an organization and comprise a Chairperson, Secretary, and a number of technical and administrative staff members.

Members from each operational area that significantly affects energy use should be represented on the team which normally would include the following:

- Operations and maintenance
- Facility management
- Purchasing

- Utilities
- Engineering
- Finance
- Environment, health, and safety
- Project management

The structure of a typical energy management team is shown in Fig. 1.5.

The energy management team is usually responsible for the following:

- Preparing energy policy
- Assessing energy management performance
- Identifying organizational strengths and weaknesses
- Approving working procedures
- Approving EnPIs
- Disseminating information
- Monitoring the implementation
- Preparing and approving energy targets and plans
- Reviewing energy policy, plans, and targets

The roles and responsibilities of team members can be expressed in a matrix as shown in Table 1.1.

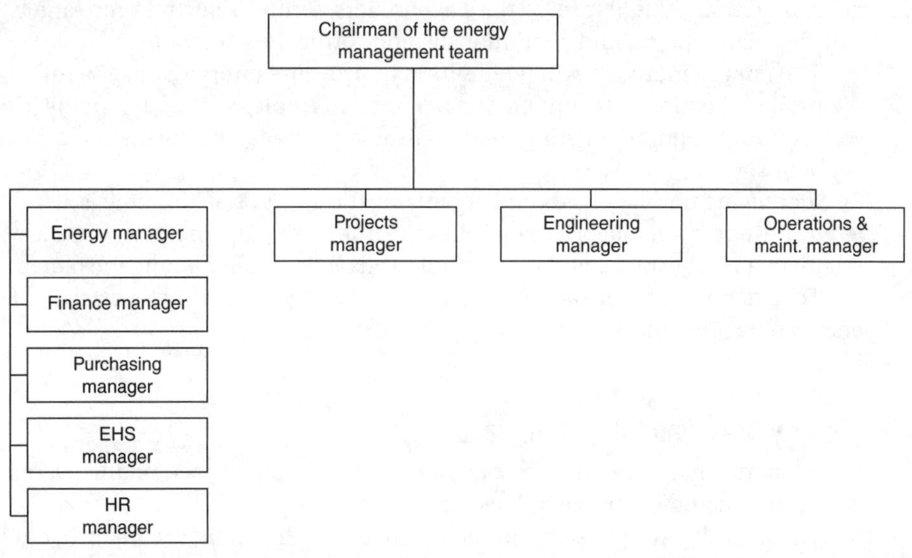

FIGURE 1.5   Structure of a typical energy management team.

	Energy management team	Engineering department	Projects department	Finance department	Purchasing department
Establishing policy, goals, and procedures	√				
Establishing EnPIs	√	√			
Planning energy projects	√	√	√		
Implementing projects	√	√	√		√
Assessing performance	√				
Financing of projects	√			√	
Review	√				

TABLE 1.1   Typical Role Matrix

1.6 Energy Manager

The energy manager can be considered the most important person on an energy management team. This person will normally be appointed by the energy management team and will be responsible for driving the energy management program.

The energy manager will be responsible for preparing energy policy, getting the necessary commitment from all the stakeholders, setting objectives and targets, implementing action plans, and continuously monitoring the program.

The energy manager will lead efforts to promote energy policy within an organization and coordinate among departments to implement action plans. He or she will conduct regular meetings to monitor and assess performance and identify problems.

Identifying training needs and organizing the appropriate training is also normally led by the energy manager. Other duties include setting up an internal audit system to assure that established standards are followed while implementing the plans.

The energy manager should report to top management on the performance of the energy management system.

1.7 Energy Usage and Baseline Data

Energy performance of an organization can be evaluated based on the overall energy consumption and the usage by specific systems.

The overall energy usage can be evaluated by trending the historic energy use (baseline data) for individual types of energy and the total energy use, as shown

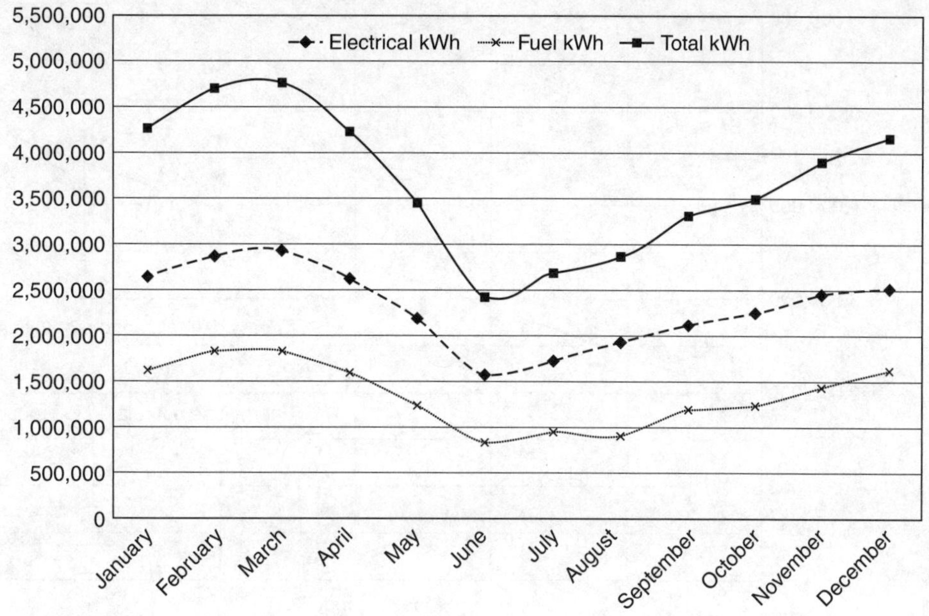

Figure 1.6   Total energy usage profile.

in Fig. 1.6. The baseline data should be for a minimum of 1 year to account for possible seasonal variations in energy use. Ideally, baseline data should be collected for a period as long as possible to ensure that all possible variations in usage are included in the analysis.

Fuel consumption may be metered in different units such as liters or cubic meters and therefore would need to be converted to a common energy unit such as kWh so that the total energy use can be computed. Conversion factors for some common fuels are summarized in Table 1.2.

The effectiveness of energy use can also be evaluated by computing an appropriate energy efficiency index (EEI). The most commonly used EEI represents the specific energy consumption (SEC) and can be defined as follows:

SEC is energy use in a particular period divided by the product or service output during the period.

Fuel	Conversion factor
Fuel oil	12 kWh/L
Diesel oil	11 kWh/L
Natural gas	11.13 kWh/m^3
LPG	7 kWh/L

Table 1.2   Typical Conversion Factors for Common Fuels

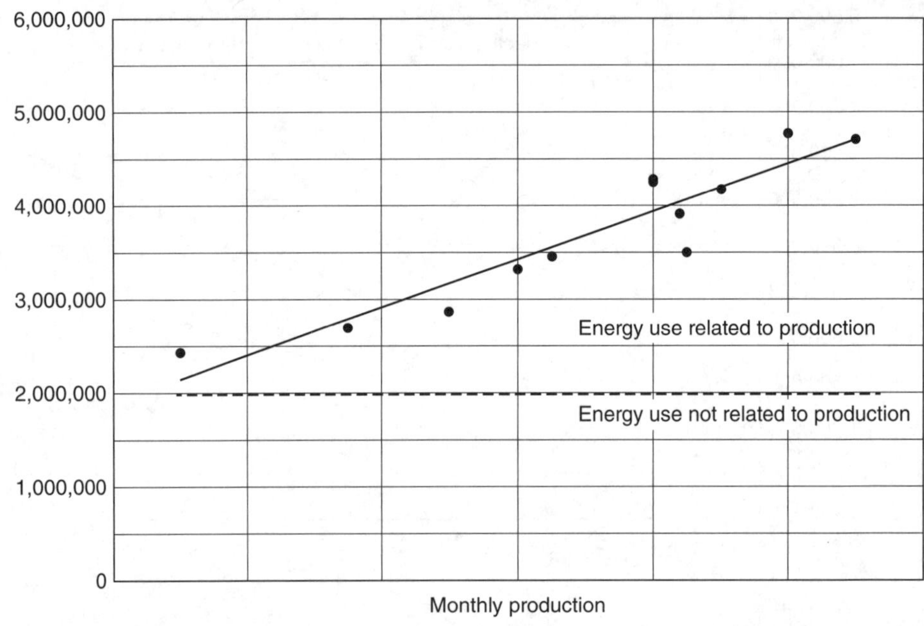

FIGURE 1.7   Typical plot of energy usage vs. output.

To compute the above index, the energy use and the total output of the organization have to be tracked. The index can be computed based on the total energy use or based on specific types of energy input (Fig. 1.6).

The product output will depend on the type of industry and can be tons of product or a number of units produced by a food manufacturing plant, number of bottles or cans produced for a beverage manufacturing plant, or the number of rooms occupied for a hotel.

Once the relevant data are compiled, they can be plotted to show how energy use varies with output (Fig. 1.7). In many industries, the energy use chart will show that there are two components of energy use, one that is dependent on the output (consumption by production machinery) and the other that is independent (utility energy consumption, lighting, and office energy use).

The SEC data can also be plotted to establish how the particular index varies with output. Figures 1.8 and 1.9 illustrate typical plots for manufacturing plants, hotels, and hospitals.

1.8   Energy Accounting Centers and EnPIs

In an effective energy management program, the main energy consumers need to be identified so that an energy balance can be performed to account for the total energy usage. An example of an energy-usage diagram is shown in Fig. 1.10.

The energy consumed by these main users must be monitored so that they can be made accountable for their respective usage. These key users which are monitored are normally called energy accounting centers (EACs). EACs can be individual machines,

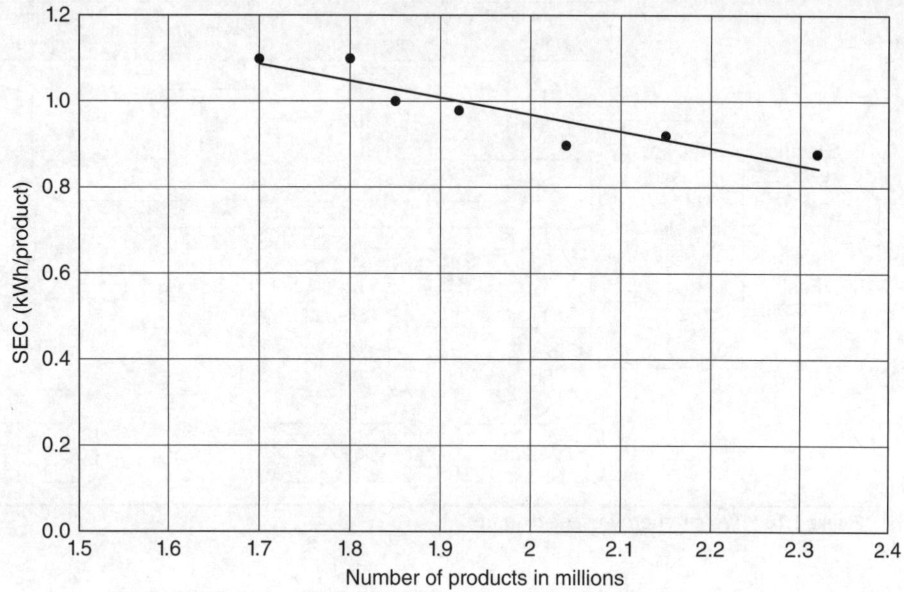

Figure 1.8 Typical plot of SEC for a manufacturing plant.

individual buildings, or departments. One possible grouping of EACs for the facility with the energy-usage breakdown illustrated earlier is shown in Fig. 1.11.

Once the EACs are identified, in addition to monitoring their overall energy usage, it is essential to monitor various key performance indicators, called energy

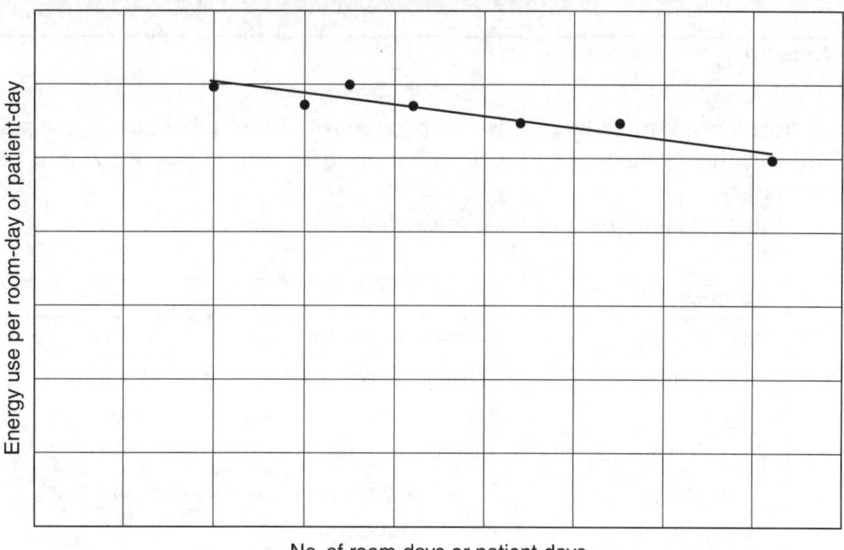

Figure 1.9 Typical plot of SEC for a hotel or hospital.

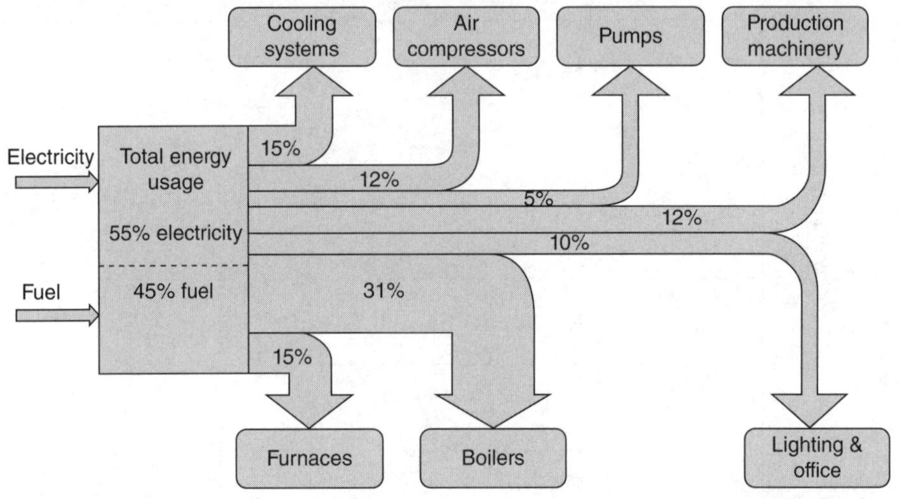

Figure 1.10 Typical energy-usage diagram.

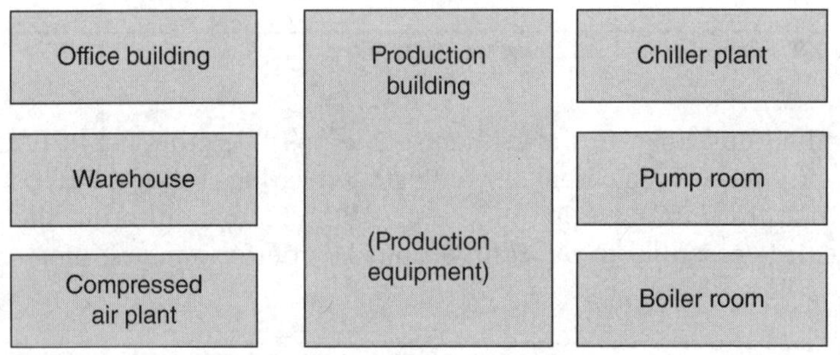

Figure 1.11 Typical arrangement of EACs.

performance indicators (EnPIs), associated with these EACs so that the performance of individual systems can be closely monitored and corrective action taken when necessary.

Table 1.3 contains some examples of EnPIs.

EAC/Area	EnPIs
Office building	Electricity usage/floor area
Chiller plant	Chiller system efficiency (COP or kW/RT)
Boiler plant	Boiler efficiency or energy usage per ton of steam produced
Compressed air plant	Specific energy (kW/100 cfm)
Production plant	Specific energy (energy usage/production output)
Warehouse	Power density (W/m^2)
Pump room	Specific energy (W/CMH)

Table 1.3 Typical EnPIs

System	Parameter
Air compressors	Specific power = kW/m^3/min
Air conditioning chillers	Coefficient of performance = kW$_{\text{cooling output}}$/kW$_{\text{energy input}}$
Refrigeration systems	Coefficient of performance = kW$_{\text{cooling output}}$/kW$_{\text{energy input}}$
Ovens and furnaces	Efficiency = heat output/heat input
Boilers	Efficiency = heat output/energy input
Fans	Specific power = fan power/airflow rate
Pumps	Pump specific power = pump power/liquid flow rate
Packaging machinery	Specific power = number of products/energy input

TABLE 1.4   List of Suitable EnPIs for Example 1.1

Example 1.1   The overall energy-usage breakdown for an industrial facility is shown in Fig. 1.18 (on page 18). Provide a list of suitable EnPIs to monitor the energy consumption of the facility.

Solution   A list of suitable EnPIs that can be used to monitor the energy usage of the facility is listed in Table 1.4.   ▲

1.9   Targets and Action Plans

Once the energy management team and the energy manager have been appointed, targets for the energy management program will need to be set. These targets are usually annual but normally set a few years in advance due to the time needed for planning and execution.

Energy objectives and targets should be measurable and a time frame set for achieving them. Objectives and targets should be consistent with the energy policy, commitments made to improve energy performance, and compliance of statutory and other requirements.

The targets have to be in line with the company energy policy and should specify targets for the EEI or SEC of the company as well as individual EnPIs set for the energy accounting centers. The targets can be in energy savings (kWh) or in terms of other parameters such as carbon emissions.

The targets set must be realistic and achievable while not being very easy to achieve to ensure maximizing the benefits of the energy management program. The set targets should be checked to ensure that they represent a significant reduction in energy use for the organization. For example, a 500,000 kWh reduction in annual energy consumption may represent a 10% savings for one organization while it may be less than 1% for another company.

Some of the factors that need to be considered when setting targets are the availability of funds, operational requirements and constraints, environmental impacts, safety and health issues, and the availability of other resources.

Once the targets are set, a suitable action plan needs to be formulated. The action plan should specify who would be responsible for implementing the plan and identify what resources (funding, instrument, technology, manpower) are required to implement the plan. Further, the action plan should be documented so that performance can be regularly monitored and updated.

1.10   Energy Audits

Once the targets are set for the energy management program, it is necessary to establish the baseline performance of the current system and identify potential areas for improvements. This is achieved through a systematic process called energy audits.

Energy audits can involve different levels of detail, depending on the objectives of the study. The most common classification is to categorize audits into three levels, from 1 to 3, depending on the depth of the audit. The type of audit selected depends on the objectives. If the energy management team is considering which facility, out of a group of sites, has the best potential for savings, then a Level 1 preliminary walk through the audit may be sufficient. On the other hand, if the energy management team requires an estimate of potential savings and cost for a particular facility for financial forecasting, a Level 2 energy survey and analysis may be necessary. Similarly, if the objective is to identify and implement specific savings measures, a Level 3 audit may be required.

1.10.1   Level 1 Audit

This involves the assessment of a building's energy cost and efficiency through the analysis of energy bills and a brief survey of the building. A Level 1 energy survey helps to identify and provide savings and cost analysis of low-cost or no-cost measures. It also provides a listing of potential capital improvements that merit further consideration, along with an initial judgment of potential costs and savings. The level of detail depends on the experience of the person performing the audit and on the specifications of the client paying for the audit.

1.10.2   Level 2 Audit

Level 2 audits normally do not include data logging, but may involve some "spot measurement" of parameters such as motor power, temperatures, and flow rates where necessary.

Although Level 2 studies require more resources than Level 1 studies, they are able to identify and short-list not only facilities for further study (as in Level 1 surveys), but also areas or measures within a facility that have a good potential for savings and where further study should be carried out. Therefore, Level 2 studies are useful exercises to be carried out before a detailed study so that the resources available for the detailed study can be better utilized.

Following are some of the main tasks carried out during a Level 2 audit:

- Information gathering on operating characteristics such as occupancy and operating hours
- Analysis of utility consumption (examples in Figs. 1.12–1.15)
- Breakdown of energy consumption into the different users (examples in Figs. 1.16–1.18)
- Benchmarking the performance of individual systems
- Identifying areas for improvements

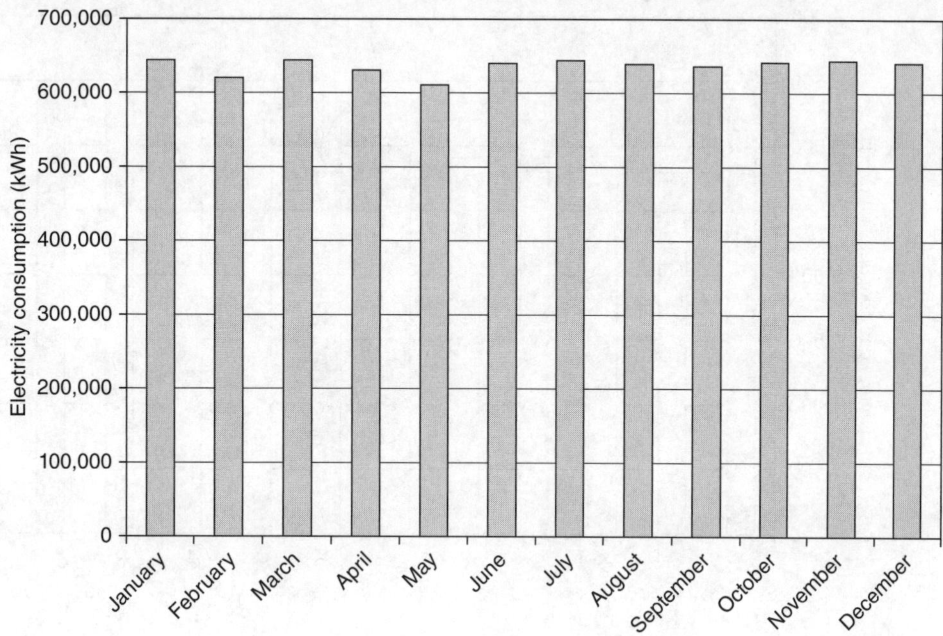

FIGURE 1.12   Electricity-consumption data.

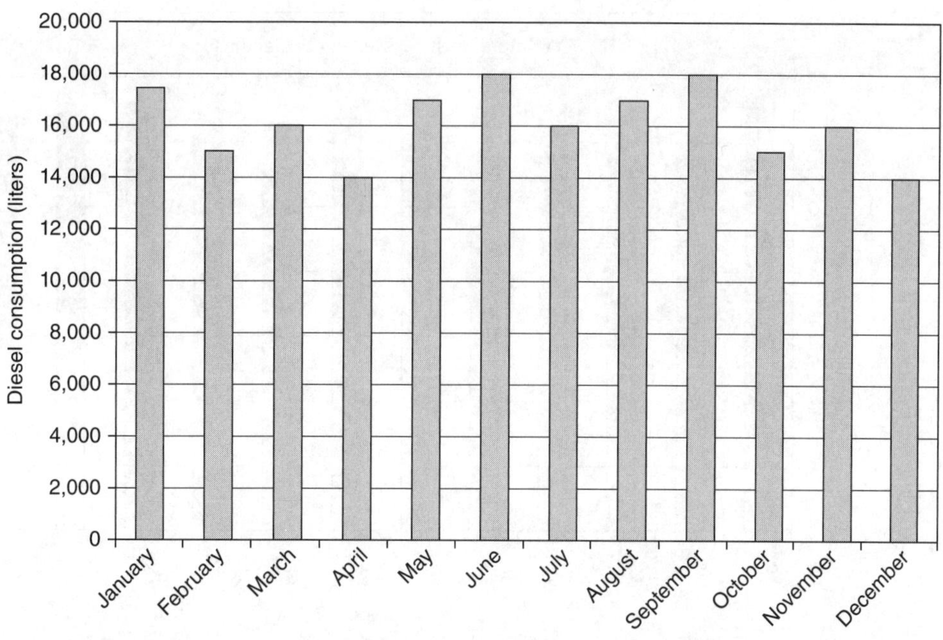

FIGURE 1.13   Diesel-consumption data.

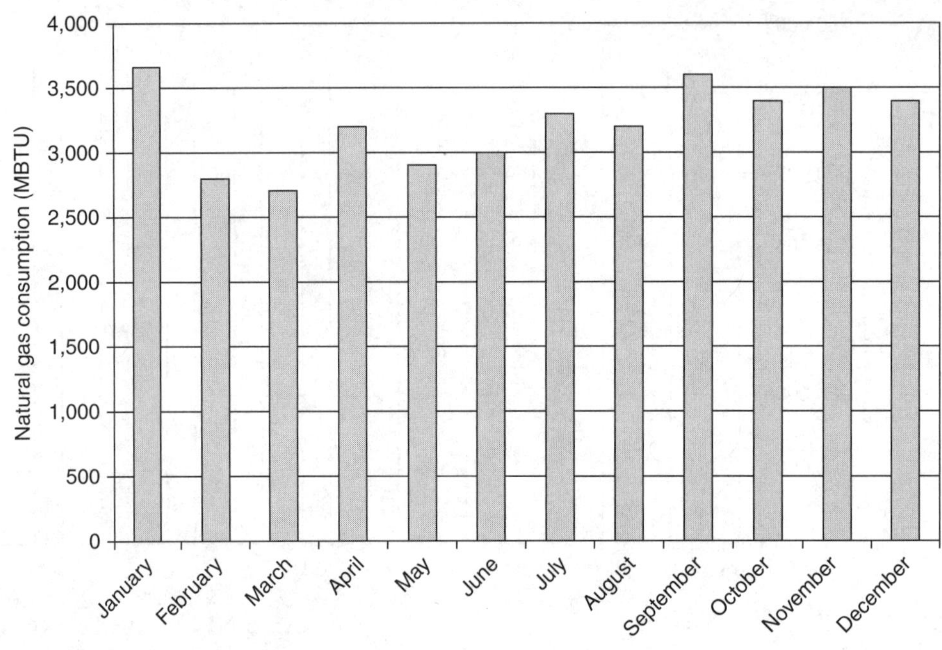

FIGURE 1.14   Natural-gas-consumption data.

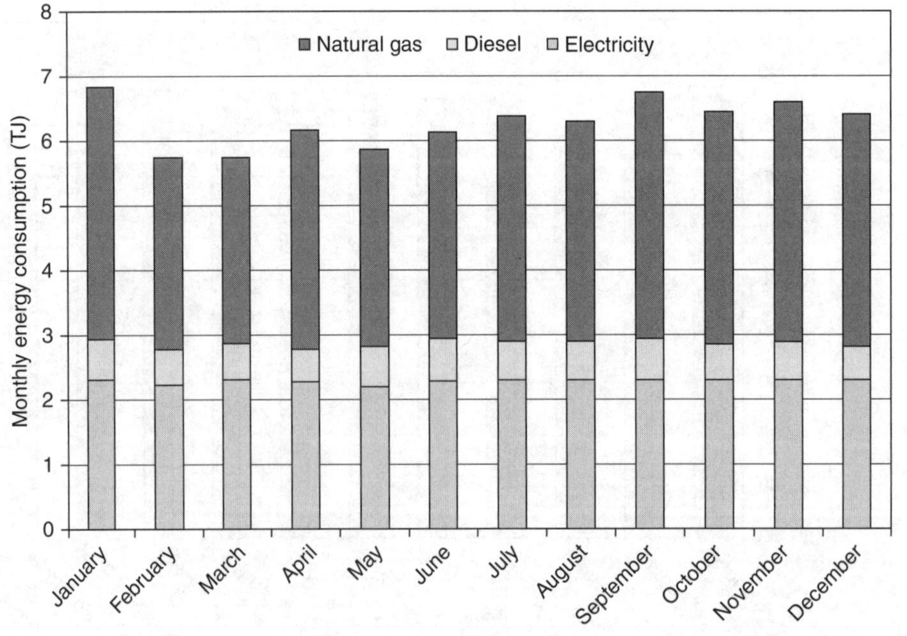

FIGURE 1.15   Total energy usage data.

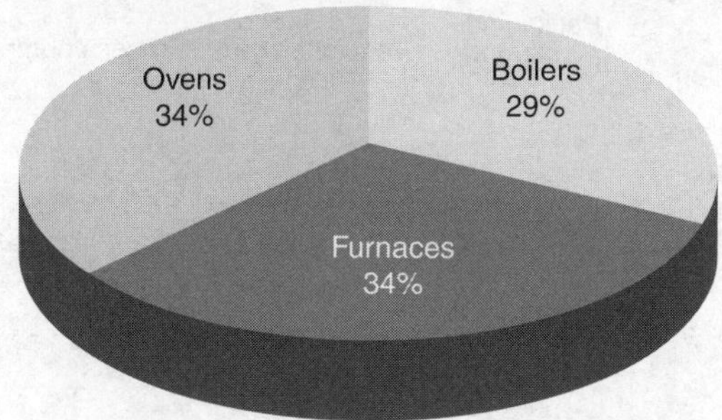

FIGURE 1.16   End-user breakdown for natural gas.

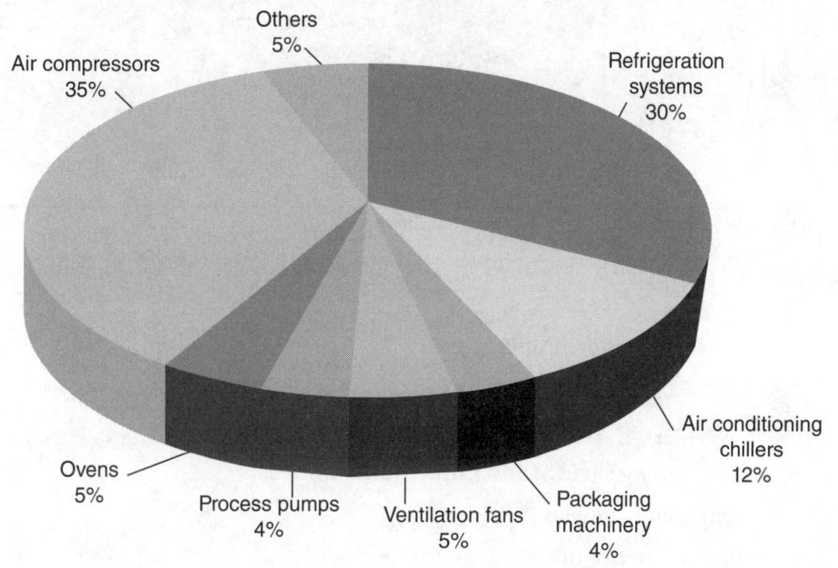

FIGURE 1.17   End-user breakdown for electricity.

1.10.3   Level 3 Audit

Level 3 audits focus on potential optimization and capital-intensive projects identified or short-listed during Level 2 audits and involve more detailed field data gathering and engineering analysis. They also provide detailed project cost and information on achievable savings with a high level of confidence, sufficient for major capital investment decisions. Therefore, Level 3 studies are also sometimes called investment-grade audits (IGA).

Following are some of the main tasks carried out during a Level 3 audit:

- Introductory meeting with the facility management team

- Interviewing relevant people to gather information

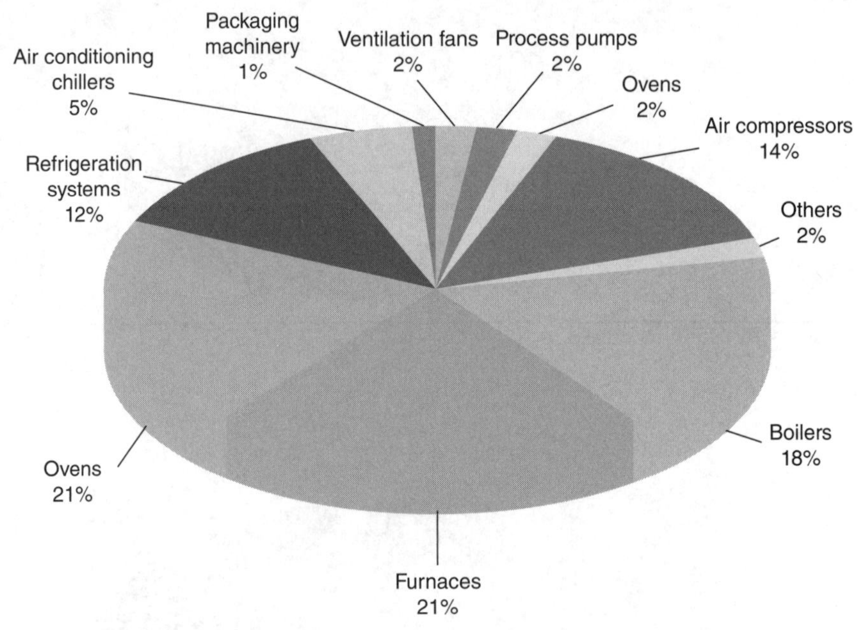

FIGURE 1.18   Overall energy end-user breakdown.

- Data collection and logging
- Data analysis
- Identifying energy-saving measures
- Estimating savings and implementation costs
- Financial analysis
- Recommendations and action plan

One of the main features of a Level 3 audit is in-depth data collection. Some typical collected data is listed in Table 1.5.

1.11   Training

For a successful implementation of an energy management program, all concerned stakeholders need to be trained on the energy management system, documentation, procedures, and internal audit requirements.

In addition, various system owners may need technical training on their specific systems so that they can better manage the respective system to perform at optimum efficiency. Examples of such systems that may require specific training for the system owners are boilers and steam generation equipment, compressed air systems, and combined heat and power (CHP) plants.

System	Parameter
Air compressor plant	• Compressed airflow rate • System pressure • Compressor power • Dryer power
Boiler plant	• Steam flow rate • Feedwater flow rate • Feedwater temperature • Steam pressure • Fuel usage • Composition of flue gas
Chiller plant	• Chilled water flow rate • Chilled water supply and return temperatures • Condenser water flow rate • Condenser water supply and return temperatures • Chiller power
Pumping systems	• Liquid flow rate • Pump power • Suction and discharge pressures
Cooling towers	• Water flow rate • Fan power • Water return and supply temperatures
Fans	• Airflow rate • Fan power • Suction and discharge pressures
Dryers and furnaces	• Fuel usage • Raw material or product throughput • Internal temperature • Exhaust temperature • Surface temperature
Process equipment	• Energy usage • Raw material or product throughput
Lighting	• Lighting power • Illuminance levels

TABLE 1.5 Typical Data Collection in Level 3 Audits

1.12 Awareness

All personnel involved in the energy management program should be fully aware of the energy policy and procedures to ensure conformity. They should also be made aware of their roles and responsibilities, significant energy users associated with their work areas, and the benefits of energy management.

In addition, the policies and plans of an energy management program need to be communicated throughout the organization. Communications can be in the form of newsletters, posters, and seminars.

1.13   Documentation

A good energy management system needs to be documented. The degree of documentation can vary depending on the size of the organization, type of activities, quantity of energy use, and complexity of processes and their interactions.

Some of the items that need to be documented include the energy policy, targets and action plans, records identified by the organization as necessary to ensure effective planning, and operation and control of processes and equipment related to its identified significant energy users.

1.14   Internal Audit

Internal audits are the main "Check" activity in the PDAC process. They are carried out periodically, normally once a year. The objective of an internal audit is to assess the performance of the energy management system to ensure that it operates in accordance with the established procedures.

The internal audit will review the different aspects of the energy management system such as the policies, EnPIs, working procedures, compliance with regulations, and other requirements. The internal audit will identify any noncompliance and will recommend corrective actions to be taken. Once the internal audit is concluded, the findings are documented and communicated to the senior management.

Internal audits are usually performed by employees of the organization who are trained on the procedures and objectives of the internal audit. For the internal audit to be effective, the persons performing the audits should be qualified, objective, and independent of the area of organization that is audited.

1.15   Management Review

The management review is the final step in the PDCA process. It is a periodic review undertaken by top management to assess the performance of the energy management system to ensure it is appropriate for the organization, able to meet the organizational needs, and that it can facilitate continual improvement.

The review should assess the energy policy and energy performance of the organization and the extent to which the energy objectives and targets have been met. Once completed, the management review may recommend changes to the energy policy, targets, or other elements of the energy management system to ensure that it is consistent with the organization's commitment to continual improvement.

CHAPTER **2**

Motors and Drives

2.1 Introduction to Motors

In industrial plants, motors used for pumps, fans, compressors, refrigeration systems, and material handling systems account for a significant portion of the total electricity consumption. In general, motors account for about 60% to 70% of the total electricity consumption in most industrial plants.

Electrical motors are devices which convert electrical energy into mechanical work. They mainly consist of a stator (stationery winding), rotor (rotating winding), enclosure, and mechanical bearings. Motors operate on the principle that "like" poles (north–north or south–south) of magnets repel, while opposite poles (north–south) attract.

When a current carrying conductor is placed in a magnetic field, a magnetic field is produced around the conductor and a force is exerted, as shown in Fig. 2.1. The direction of the force, from "Fleming's right-hand rule," states that when the thumb and the first two fingers of the right hand are placed at right angles to each other and the forefinger is placed in the direction of the magnetic field, the second finger points in the direction of the current and the thumb points in the direction of the force.

If the current carrying conductor is looped as shown in Fig. 2.2, upward and downward forces will be exerted on the opposite sides of the loop, resulting in the rotation of the coil. In actual motors, the rotor consists of many coils of conductors to form magnetic poles which interact with the stator poles to provide continuous torque.

The magnetic fields in the stator and rotor of motors are generated using electromagnets or permanent magnets. The main types of motors used in industrial plants are AC (alternating current) induction motors, DC (direct current) motors, and synchronous motors.

2.1.1 AC Induction Motors

In AC induction motors, an AC current applied to the stator creates a magnetic field in the stator winding. When the conductors of the rotor cut the lines of flux, a current is induced in the rotor, creating a magnetic field which opposes the magnetic field in the stator, resulting in the rotation of the rotor. A typical AC motor is shown in Fig. 2.3.

The most common type of rotor used in AC induction motors is "squirrel cage," which has aluminum or copper rotor bars joined at the ends by end rings (Fig. 2.4). The rotating magnetic field in the stator induces a voltage in the rotor bars, which causes a current to flow through them. The rotor currents then generate a magnetic field, which

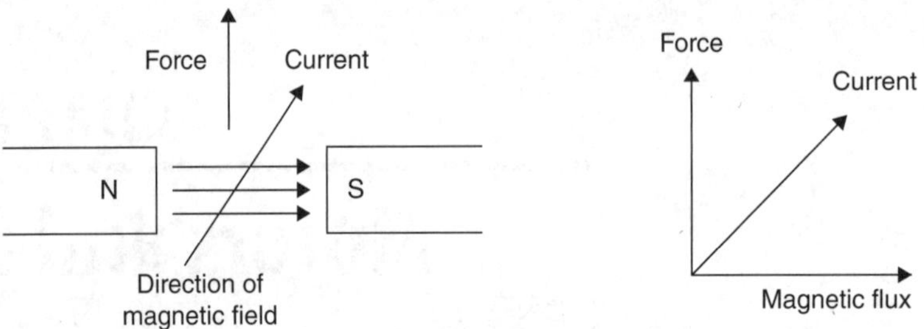

FIGURE 2.1   Direction of the force on a current-carrying conductor in a magnetic field.

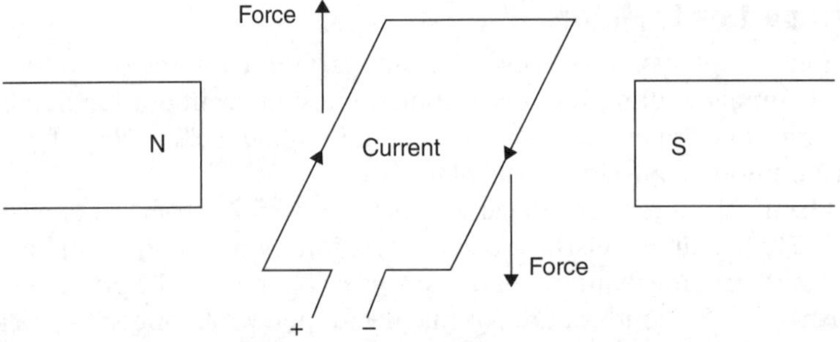

FIGURE 2.2   Direction of forces on a current-carrying loop placed in a magnetic field.

FIGURE 2.3   AC induction motor (courtesy of ABB).

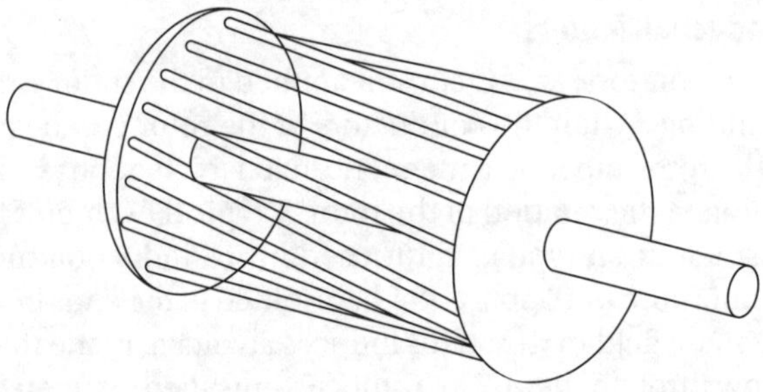

FIGURE 2.4   Squirrel cage rotor.

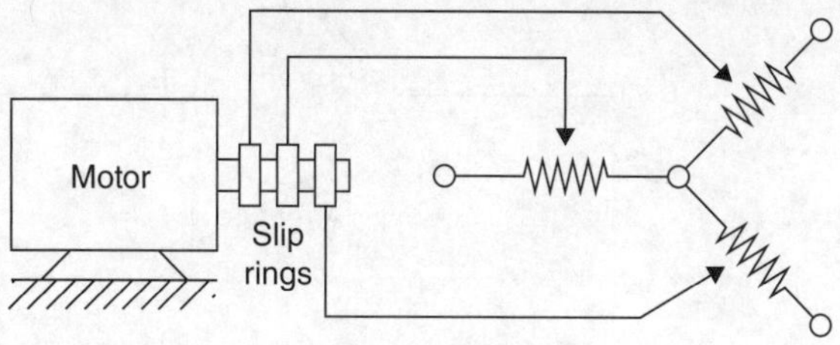

FIGURE 2.5 Arrangement of a wound rotor with slip rings.

interacts with the field in the stator. The torque developed by the rotor depends on the resistance of the rotor. A high resistance results in a high starting torque and vice versa.

The other common type of rotor is the "wound rotor" where wires are wound into rotor slots and are connected to external resistors through slip rings, as shown in Fig. 2.5. The resistors are variable and are used to alter the rotor resistance and thereby vary motor speed and torque.

Squirrel cage rotors are generally preferred due to lower cost, whereas wound rotor motors are used for applications which require a higher starting torque and speed variation.

2.1.2 DC Motors

In DC motors, direct current is passed through a rotor wire placed between wound or permanent magnets of the stator. DC current is supplied to the rotor (while it is rotating) through a set of contacts, called a commutator, which are in contact with stationary conductors called brushes. The electricity flow through the rotor coil creates a magnetic field. The direction of the flow of current is changed twice every cycle, which changes the polarity of the rotor electromagnet. The rotor magnetic field interacts with the wound or permanent magnet of the stator, which makes the rotor spin. A typical DC motor is shown in Fig. 2.6.

FIGURE 2.6 Arrangement of a DC motor (courtesy of ABB).

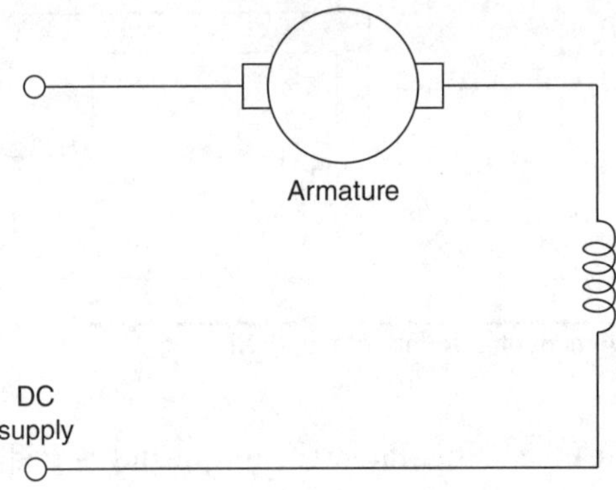

FIGURE 2.7   Arrangement of series-wound DC motors.

DC motors are generally more expensive than AC induction motors because of the need for a commutator, brushes, and rotor winding. They are generally used only for special applications such as lifts and cranes, which require precise speed and torque control.

DC motors are generally classified as series wound, shunt wound, or compound wound, based on the connection between the armature (rotor) and field (stator) winding.

Series Wound

In series-wound DC motors, current flows in series through both the field and the armature windings as shown in Fig. 2.7. Such motors have low-resistance field and armature circuits.

When the starting voltage is applied, the high current in the winding results in a strong magnetic field and therefore a high starting torque.

Series-wound DC motors are used for applications such as cranes and hoists, which have high starting loads. As the speed–torque characteristic curve in Fig. 2.8 shows,

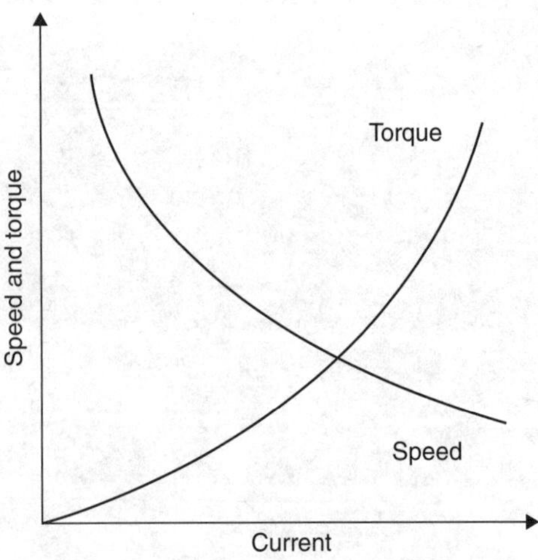

FIGURE 2.8   Speed–torque characteristic for series-wound DC motors.

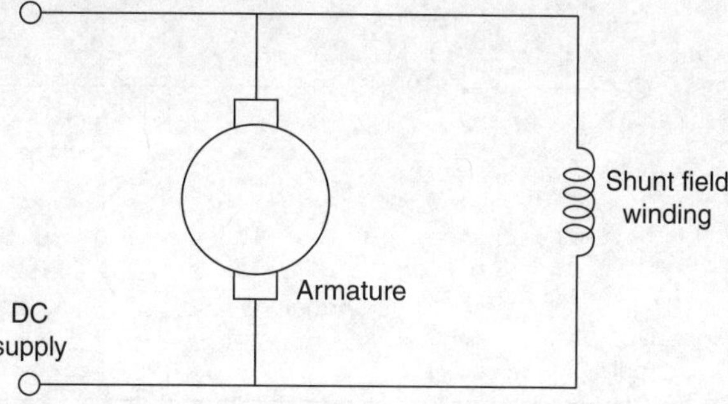

FIGURE 2.9 Arrangement of shunt-wound DC motors.

when the torque increases, the motor speed drops. Therefore, this type of motor cannot be used for applications which require constant speed operation under variable load conditions.

Shunt Wound

In shunt-wound DC motors, the field and armature windings are connected in parallel as shown in Fig. 2.9. The shunt field winding is made up of many more turns of a small-gauge wire to increase its resistance when compared to the series-field winding. As a result, the current flow through the field winding is lower.

Since the power supply is connected directly to the field winding, the field current is constant and the torque developed by the motor depends on the armature current. Therefore, at start-up, the torque is low but reaches its maximum at full speed. These motors are used for applications such as conveyors where despite variations in torque, a relatively constant operating speed is required. The speed-torque characteristic curve for a shunt-wound motor is shown in Fig. 2.10.

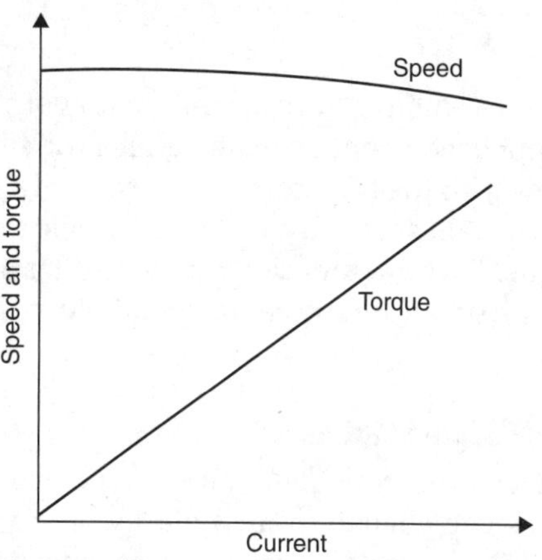

FIGURE 2.10 Speed–torque characteristic for shunt-wound DC motors.

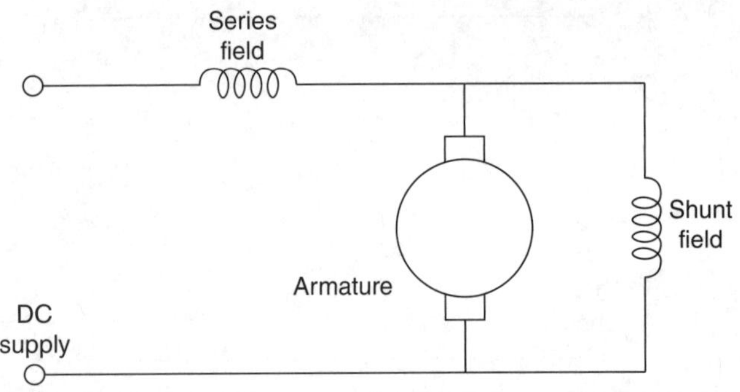

Figure 2.11 Arrangement of compound-wound DC motors.

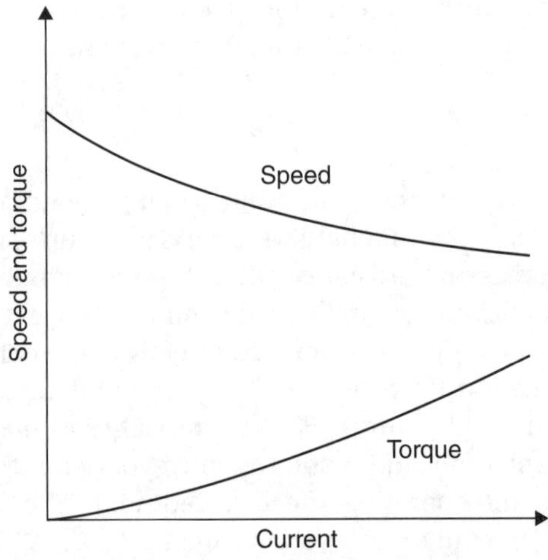

Figure 2.12 Speed–torque characteristic for compound-wound DC motors.

Compound Wound

Compound-wound DC motors have a combination of series and parallel wound field windings. One shunt field is connected in parallel with the armature, while the other is connected in series with the armature as shown in Fig. 2.11. The shunt field provides the advantage of constant speed, while the series field provides the ability to develop a high starting torque. Such motors can handle sudden increase in load without great change in speed. The speed-torque characteristic curve for a compound-wound motor is shown in Fig. 2.12.

2.1.3 Permanent-Magnet Motors

Permanent-magnet motors consist of a permanent magnet stator and a wound rotor as shown in Fig. 2.13. The change in the polarity of the magnetic field in the rotor is achieved by switching current between the coils using a commutator and brushes. These motors have a good starting torque and lower motor losses.

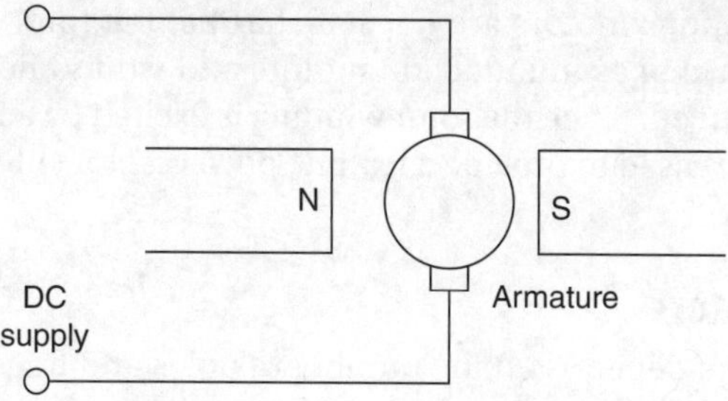

FIGURE 2.13   Arrangement of permanent magnet DC motors.

2.1.4   Synchronous Motors

Synchronous motors have stator windings where an AC voltage is applied to produce a rotating magnetic field. A DC voltage is supplied to the rotor via brushes to create a magnetic field, which interacts with the rotating magnetic field in the stator, resulting in the rotation of the rotor. The arrangement and winding configuration of a synchronous motor are shown in Figs. 2.14 and 2.15. At steady state, the speed of the rotor is the same as the speed of the rotating magnetic field in the stator. As the rotor does not rely on magnetic induction from the stator as in the case of AC induction motors, slip is not required to produce torque. Therefore, motor speed is independent of load. Synchronous motors are used when a precise constant speed is required.

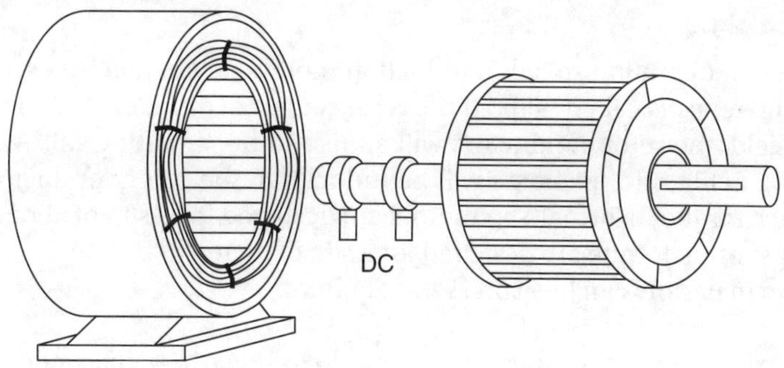

FIGURE 2.14   Arrangement of a synchronous motor.

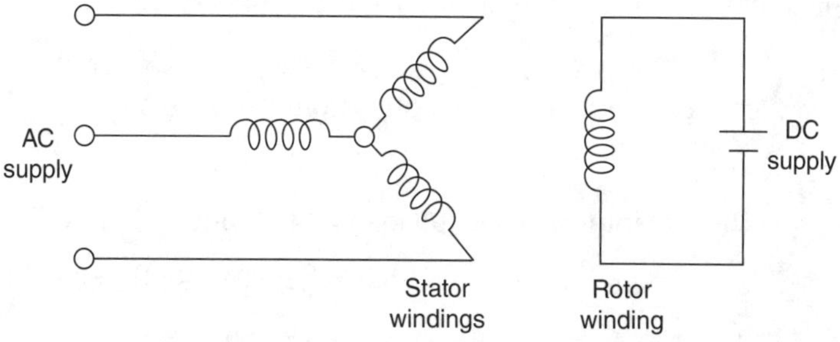

FIGURE 2.15   Arrangement of windings in a synchronous motor.

As synchronous motors are not self-starting, the rotor consists of a squirrel cage winding for starting as in induction motors and wound field poles for operation at a synchronous speed. Since the rotor winding is excited by a DC supply, the field excitation can be increased to provide a leading power factor (PF).

2.2   Speed of Motors

Speed of motors depends on the number of poles and frequency of the power supply and can be expressed as follows:

$$\text{Synchronous speed (rpm)} = (120 \times \text{power supply frequency in Hz})/(\text{number of poles in the motor})$$

Example 2.1   Compute the synchronous speed for a four-pole motor operating on 50 Hz power supply.

Solution

$$\text{Synchronous speed} = (120 \times 50)/4$$

$$= 1500 \text{ rpm} \quad \blacktriangle$$

2.3   Slip in Motors

The actual operating speed of induction motors is less than the synchronous speed. The difference is called "slip." If the rotor rotates at the same speed as the stator magnetic field, the rotor conductors will appear to be "standing still" with respect to the rotating field and no voltage will be induced in the rotor conductors (no current to produce torque). Normal-slip motors are designed for a slip of about 5%. Motors with a high starting torque are designed for a slip of up to 20%.

Slip in motors can be expressed as follows:

$$\text{Slip (\%)} = [(\text{synchronous speed} - \text{full-load speed})/(\text{synchronous speed})] \times 100$$

Example 2.2   Compute the percentage of slip for a four-pole motor operating on 50 Hz power supply when the full-load motor speed is 1440 rpm.

Solution
Synchronous speed for a four-pole motor (from Example 2.1)

$$= 1500 \text{ rpm}$$

$$\text{Motor speed at full load} = 1440 \text{ rpm}$$

$$\text{Slip} = [(1500 - 1440)/1500] \times 100$$

$$= 4\% \quad \blacktriangle$$

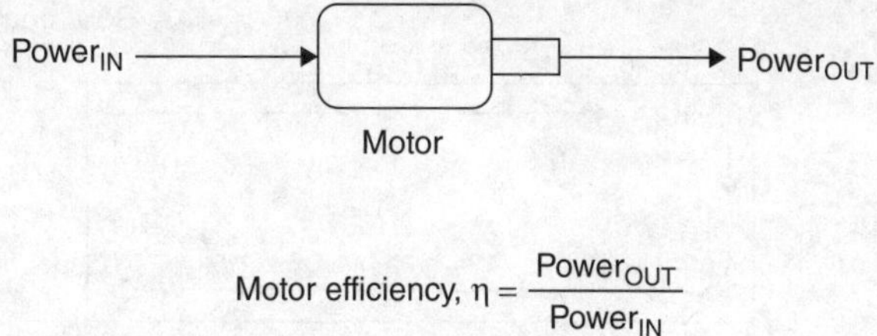

$$\text{Motor efficiency, } \eta = \frac{\text{Power}_{OUT}}{\text{Power}_{IN}}$$

FIGURE 2.16 Definition of motor efficiency.

2.4 Motor Efficiency

Efficiency of a motor is a measure of how well it can convert the input power into useful work. Some devices such as electric heaters can convert 100% of the power consumed into heat. However, in other devices such as motors, the total energy consumed cannot be fully converted into usable energy as a certain portion is lost and is not recoverable, because it is expended as losses associated with operating the device (Fig. 2.16). Therefore, it is necessary to provide more than 1 kW of electrical power to produce 1 kW of mechanical output.

2.5 Losses in Motors

In motors, the energy that is lost is emitted in the form of heat. As shown in Fig. 2.17, motor losses can be categorized as copper losses, iron (core) losses, friction and windage losses, and stray losses.

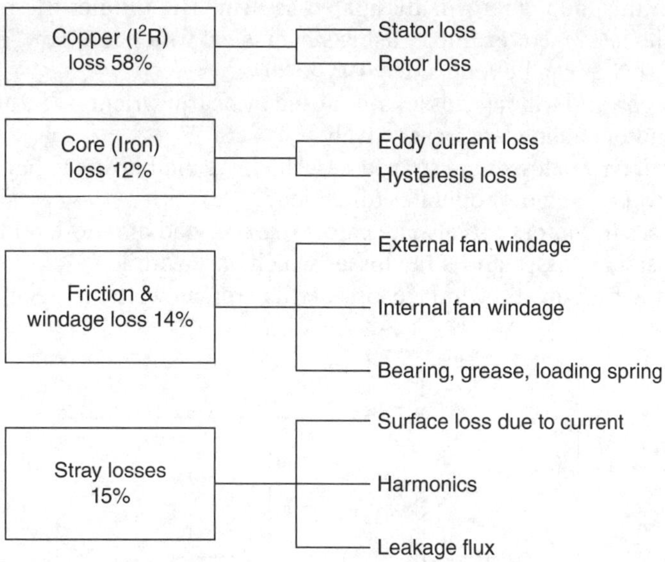

FIGURE 2.17 Losses in a typical motor.

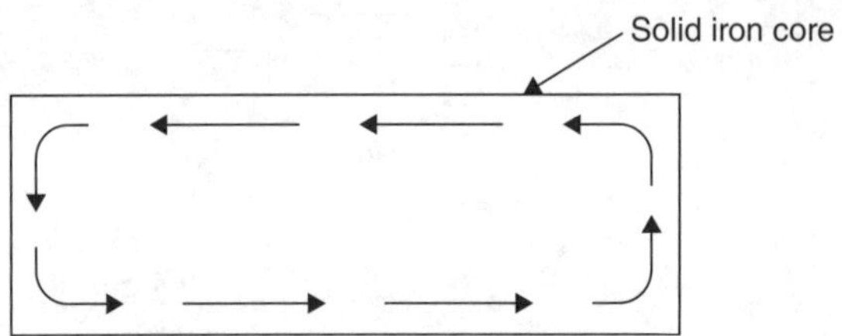

FIGURE 2.18   Eddy current in a solid iron core.

The largest single loss in a motor is the resistance loss (I^2R), which accounts for more than half of the losses. Resistance loss occurs when current passes through the stator windings, rotor conductors, and end rings and is dissipated as heat.

The core losses are due to eddy currents and hysteresis. The magnetic field of the stator while inducing a voltage in the rotor winding also induces a voltage in the iron core, resulting in current flows in the iron core (Fig. 2.18). These current flows are called eddy currents and are dissipated as heat leading to losses. To minimize eddy current losses, the core is made up of layers called laminations (sheets of steel) with insulation in between to confine eddy currents within each lamination (Fig. 2.19).

Hysteresis loss is due to residual magnetism in electromagnets. When a magnetic field is applied and later removed, the magnetic flux density does not reach zero and a further magnetic field needs to be applied in the opposite direction to bring the residual magnetism to zero. The path followed during the magnetization cycle is called the hysteresis loop. Materials that have high magnetic retention are called "hard" magnetic materials, while those having lower magnetic retention are termed "soft" magnetic materials. Typical hysteresis loops for "hard" and "soft" magnetic materials are shown in Fig. 2.20. When a magnetic material is subjected to continuous cycling through this loop, energy is dissipated as heat. The thinner the hysteresis loop, the lower the losses. For example, using silicon steel which is a "soft" magnetic material, the core losses can be reduced by 10% to 25%.

Friction and windage losses are caused by bearing friction and air resistance (drag) of the motor cooling fans, respectively.

Other type of losses are stray losses which are mainly due to harmonic energy generated when the motor operates under load.

Losses in motors can also be categorized as load and no-load losses. Load losses are resistance losses and stray losses which vary with load. No-load losses are core losses and friction and windage losses which remain constant regardless of the load.

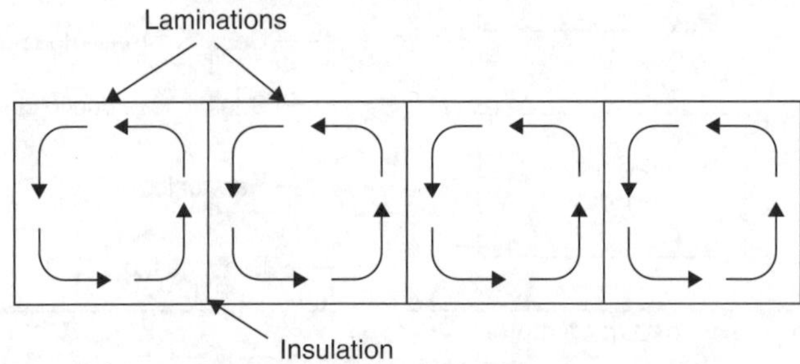

FIGURE 2.19   Eddy currents in a laminated iron core.

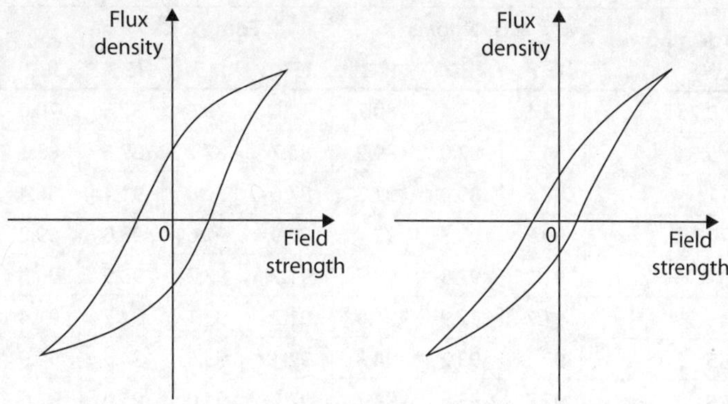

FIGURE 2.20   Hysteresis loops for "hard" and "soft" magnetic materials.

2.6   High-Efficiency Motors

The efficiency of motors depends on the size and normally ranges from about 78% to 93% for standard efficiency motors. In addition to these standard motors, various premium-efficiency motors which operate at efficiency of about 3% to 7% higher than the standard designs are also available.

In the USA, the National Electrical Manufacturers Association (NEMA) has set minimum ratings for standard- and premium-efficiency motors. The NEMA premium-efficiency motors are much more efficient than standard-efficiency motors. They are also more efficient than the EPACT (Energy Policy Act of 1992) compliant motors which are required to comply with ASHRAE Standard 90.1, "Energy standard for buildings except low-rise residential buildings."

Similarly, the CEMEP (European Committee of Manufacturers of Electrical Machines and Power Electronics) and the European Union (EU) have a motor efficiency classification: EFF1, EFF2, and EFF3 (EFF1 being the highest efficiency).

The International Electrotechnical Commission (IEC) also has introduced a new standard (IEC 60034-30) relating to energy-efficient motors. This standard defines new efficiency classes for motors and harmonizes the current different requirements for induction motor efficiency levels around the world.

IEC 60034-30 defines three IE (International Efficiency) classes: IE 1 (standard efficiency), IE 2 (high efficiency), and IE 3 (premium efficiency). Table 2.1 shows a comparison of the IE classes with the NEMA, EPACT, and EU ratings.

The minimum specified efficiency levels for IE1, IE 2, and IE 3 motors operating at 50 and 60 Hz for a selected range of motor sizes are listed in Tables 2.2 and 2.3.

Motor type	IEC 60034-30	NEMA, EPACT & EU
Standard	IE 1	Standard, EFF2
High efficiency	IE 2	EPACT, EFF1
Premium efficiency	IE 3	NEMA Premium

TABLE 2.1   Comparison Between Standards

Motor power (kW)	Two poles			Four poles			Six poles		
	IE 1	IE 2	IE 3	IE 1	IE 2	IE 3	IE 1	IE 2	IE 3
0.75	72.1	77.4	80.7	72.1	79.6	82.5	70.0	75.9	78.9
5.5	84.7	87.0	89.2	84.7	87.7	89.6	83.1	86.0	88.0
11	87.6	89.4	91.2	87.6	89.8	91.4	86.4	88.7	90.3
22	89.9	91.3	92.7	89.9	91.6	93.0	89.2	90.9	92.2
37	91.2	92.5	93.7	91.2	92.7	93.9	90.8	92.2	93.3
45	91.7	92.9	94.0	91.7	93.1	94.2	91.4	92.7	93.7
55	92.1	93.2	94.3	92.1	93.5	94.6	91.9	93.1	94.1
75	92.7	93.8	94.7	92.7	94.0	95.0	92.6	93.7	94.6
110	93.3	94.3	95.2	93.3	94.5	95.4	93.3	94.3	95.1

TABLE 2.2 Comparison Between IE 1, IE 2, and IE 3 Motors Operating at 50 Hz

Motor power (kW)	Two poles			Four poles			Six poles		
	IE 1	IE 2	IE 3	IE 1	IE 2	IE 3	IE 1	IE 2	IE 3
0.75	77.0	75.5	77.0	78.0	82.5	85.5	73.0	80.0	82.5
5.5	86.0	88.5	89.5	87.0	89.5	91.7	85.0	89.5	91.0
11	87.5	90.2	91.0	88.5	91.0	92.4	89.0	90.2	91.7
22	89.5	91.0	91.7	91.0	92.4	93.6	91.0	91.7	93.0
37	91.5	92.4	93.0	92.4	93.0	94.5	91.7	93.0	94.1
45	91.7	93.0	93.6	93.0	93.6	95.0	91.7	93.6	94.5
55	92.4	93.0	93.6	93.1	94.1	95.4	93.0	93.6	94.5
75	93.0	93.6	94.1	93.2	94.5	95.4	93.0	94.1	95.0
110	93.0	94.5	95.0	93.5	95.0	95.8	94.1	95.0	95.8

TABLE 2.3 Comparison Between IE 1, IE 2, and IE 3 Motors Operating at 60 Hz

High-efficiency motors are designed to minimize losses by using various techniques and are summarized in Table 2.4.

2.7 Impact of Motor Loading on Efficiency

The operating efficiency of motors also depends on the loading. As Fig. 2.21 shows, motor efficiency is close to its full-load efficiency when loaded above 40%, but drops significantly if the motor is loaded lower than this value. Therefore, if a motor is loaded to less than about 40%, it should be considered for replacement with a correctly sized motor. However, care should be taken to ensure that the replacement motor meets the starting torque required for a particular application, as in some instances motors are "oversized" to overcome a high starting torque.

Type of loss	Remediation
Stator resistance losses	Using wire with lower resistance and improving stator slot design
Rotor resistance losses	Increasing size of aluminum conductor bars and end rings to reduce resistance and use of copper conductors in larger motors
Hysteresis losses	Using steel containing up to 4% silicon instead of normal carbon steel for the laminations and lengthening the core to reduce magnetic flux density
Eddy current losses	Using thinner laminations and increasing insulation between laminations
Friction losses	Bearings and seal selection to lower friction
Windage losses	Fan selection and sizing to minimize drag
Stray losses	Improving rotor slot geometry

TABLE 2.4   Techniques Used to Reduce Motor Losses

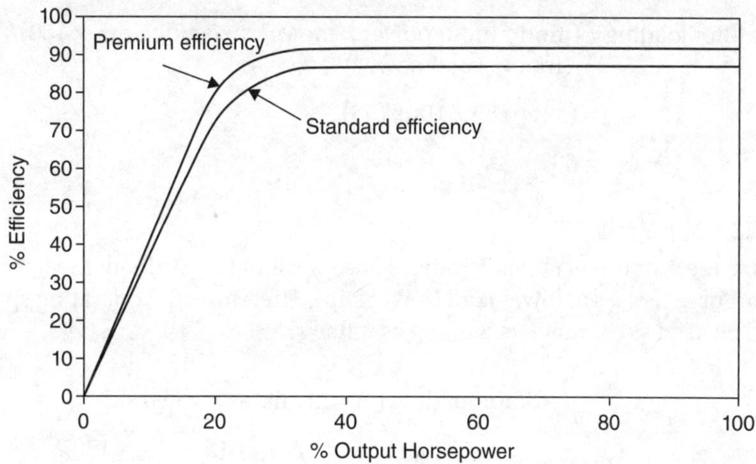

FIGURE 2.21   Motor efficiency vs. loading.

2.8   Estimating Motor Loading

Since motor operating efficiency is dependent on loading, it is necessary to know the loading of motors to ensure efficient operation. Motor capacity is rated based on the output power, and therefore motor loading is the ratio of the actual motor power output at the operating load to the rated capacity of the motor. Although it is possible to measure the output of motors under laboratory test conditions, it is not practical to do so under actual operating conditions. Therefore, the loading of motors generally needs to be estimated. Two common methods used for estimating the loading of motors are the "input method" and the "slip method."

2.8.1   Input Method

Input method involves measuring the input power to the motor and multiplying it by the rated motor efficiency to estimate the motor output power and loading as follows:

$$\text{Motor output power} = \text{motor input power} \times \text{motor rated efficiency}$$

$$\text{Motor loading (\%)} = (\text{motor input power} \times \text{motor rated efficiency} \times 100)/$$
$$(\text{motor rated power})$$

Since in this method the efficiency of the motor at the operating load is assumed to be the full-load efficiency, it may not provide a good estimation when the motor is loaded below about 40% (when the efficiency drops significantly).

Example 2.3   The input power to a 55-kW motor is measured to be 42 kW. If the design motor efficiency is 89%, estimate the motor loading.

Solution

$$\text{Motor loading} = (\text{motor input power} \times \text{motor rated efficiency} \times 100)/$$
$$(\text{motor rated power})$$

$$= (42 \times 0.89 \times 100)/(55)$$

$$= 68\% \quad \blacktriangle$$

2.8.2   Slip Method

Since the amount of slip in a motor is proportional to the load, the operating speed of a motor increases at lower load (lower slip). Therefore, the operating speed of a motor can be used to estimate its loading as follows:

$$\text{Motor loading (\%)} = \text{actual slip/rated slip}$$

$$= [(S_s - S)/(S_s - S_r)] \times 100\%$$

where S_s = synchronous speed, S = actual motor operating speed, and S_r = rated motor speed.

Example 2.4   The motor operating speed and rated full-load speed of a two-pole motor are 2920 and 2870 rpm, respectively. Estimate the motor loading using the slip method.

Solution

$$\text{Synchronous speed} = (120 \times 50)/2$$

$$= 3000 \text{ rpm}$$

$$\text{Motor loading} = [(3000 - 2920)/(3000 - 2870)] \times 100$$

$$= 61.5\% \quad \blacktriangle$$

2.9 Economics of Energy-Efficient Motors

High-efficiency motors consume less energy than standard-efficiency motors due to the lower input power required to produce the same output, resulting in lower energy cost. However, the economic viability of using high-efficiency motors instead of standard-efficiency motors depends on the cost associated with the actual application. In general, there are the following three main types of applications:

- Replacement of an existing standard-efficiency motor with a high-efficiency motor
- Replacement of a defective motor with a new high-efficiency motor
- Use of a high-efficiency motor for a new installation

The economics for the above three applications can be quite different. For the first application where an existing motor is to be replaced, the full cost of the new high-efficiency motor has to be considered. In comparison, for the third application of a new installation, only the incremental cost for the high-efficiency motor (compared to the cost of a standard-efficiency motor) needs to be considered. Similarly, for the application where a defective motor is to be replaced, the incremental cost and possible drop in efficiency due to the repair need to be taken into account.

Some of the reasons for possible drop in motor efficiency after repair are damage to laminations while stripping the motor and damage to the iron core due to high temperatures experienced during failure. Possible efficiency losses due to the poor quality of rewinding are indicated in Table 2.5.

Some of the factors that affect the economic viability of different applications are summarized with brief explanations in Table 2.6.

The following examples illustrate the economics associated with each type of application.

Example 2.5 A 22-kW motor of 88% efficiency is used for an application which requires it to operate 12 hours a day and 365 days a year. Calculate the savings that will result if this motor is replaced with a premium-efficiency motor of 93% efficiency. Assume that the motor is operating at its rated capacity. Estimate the simple payback period if the cost of the new motor is $3,000 and the average electricity tariff is $0.15/kWh.

Solution

$$\text{Motor efficiency, } \eta = \frac{\text{power}_{\text{OUT}}}{\text{power}_{\text{IN}}}$$

Therefore,

$$\text{Power}_{\text{IN}} = \frac{\text{power}_{\text{OUT}}}{\text{motor efficiency}}$$

$$\text{Estimated power input to the existing motor, } P_1 = 22/0.88 = 25 \text{ kW}$$

$$\text{Estimated power input to the proposed new motor, } P_2 = 22/0.93 = 23.7 \text{ kW}$$

$$\text{Savings in power } (P_1 - P_2) = (25 - 23.7) \text{ kW}$$

$$= 1.3 \text{ kW}$$

$$\text{Energy savings} = \text{power savings (kW)} \times \text{operating hours (hours)}$$

$$\text{Annual energy savings} = 1.3 \times 12 \times 365 \text{ kWh/year}$$

$$= 5694 \text{ kWh/year}$$

$$\text{Annual cost savings} = \text{annual energy savings in kWh} \times \text{energy tariff}$$

$$= 5694 \times 0.15$$

$$= \$854.10/\text{year}$$

$$\text{Simple payback period} = \text{cost of new motor/annual cost savings}$$

$$= \$3000/\$854.10$$

$$= 3.5 \text{ years} \quad \blacktriangle$$

Motor power (kW)	Possible reduction in motor efficiency (%)
7.5	5–6
37.3	4
75	2–3
150	<2

TABLE 2.5   Efficiency Drop Due to Poor Rewinding

Factor	Impact on economic viability
Efficiency of the existing motor	The lower the efficiency of the existing motor, the higher the energy savings.
Motor operating hours	The longer the operating hours, the higher the energy savings.
Motor loading	If the loading of the existing motor is low, it would result in poor motor operating efficiency. A correctly sized new motor would result in much better efficiency and higher savings.
Electricity tariff	The higher the tariff, the higher the savings.
Age of the existing motor	The older the existing motor, the justification to replace will be stronger due to the fact that the remaining motor life would be short and there would be a drop in efficiency due to aging.
Cost of the new high-efficiency motor	The lower the cost of the new high efficiency motor, the higher the return on investment.
Financial incentives	If financial incentives from authorities or utility companies are available to offset part of the cost of the new motor, better will be the resulting return on investment.
Drop in efficiency due to repairs	When a motor is defective and needs to be repaired, the possible drop in efficiency due to the repair and the cost of the repair need to be considered. The higher the repair cost and higher the drop in efficiency, the stronger the justification for using a higher efficiency motor.

TABLE 2.6   Factors Affecting the Economic Viability of Using High-Efficiency Motors

Example 2.6   The stator winding of a 55-kW motor which operates 18 hours a day and 250 days a year is defective. The cost of repairing the defective motor is $4,000. However, the efficiency of the motor is expected to drop by 1% from its rated efficiency of 89% after the repair. Calculate the energy and cost savings that will result if this motor is replaced with a premium-efficiency motor of 93% efficiency rather than repairing it. Assume that the motor is operating at its rated capacity. Estimate the simple payback period for replacing the defective motor with the premium-efficiency motor (93% efficiency) if the cost of the premium-efficiency motor is $10,000 and the average electricity tariff is $0.15/kWh.

Solution

$$\text{Power}_{\text{IN}} = \frac{\text{power}_{\text{OUT}}}{\text{motor efficiency}}$$

Estimated power input to the motor after repair, $P_1 = 55/(0.89 - 0.01) = 62.5$ kW

Estimated power input to the premium-efficiency motor,
$$P_2 = 55/0.93 = 59.1 \text{ kW}$$

$$\text{Savings in power } (P_1 - P_2) = (62.5 - 59.1) \text{ kW}$$

$$= 3.4 \text{ kW}$$

$$\text{Energy savings} = \text{power savings (kW)} \times \text{operating hours (hours)}$$

$$\text{Annual energy savings} = 3.4 \times 18 \times 250 \text{ kWh/year}$$

$$= 15{,}300 \text{ kWh/year}$$

$$\text{Annual cost savings} = \text{annual energy savings in kWh} \times \text{energy tariff}$$

$$= 15{,}300 \times 0.15$$

$$= \$2295/\text{year}$$

$$\text{Simple payback period} = \text{incremental cost of new motor/annual cost savings}$$

$$= (\$10{,}000 - \$4000)/\$2295$$

$$= 2.6 \text{ years} \quad \blacktriangle$$

Example 2.7   A 37-kW motor with a rated efficiency of 90% is selected for a new fan installation, which is expected to run 24 hours a day and 365 days a year. Calculate the savings in input power to the motor that can be achieved for this application if a premium-efficiency motor with 94% efficiency is used. Assume that the motor is operating at the rated capacity.

If the cost of the originally selected motor and the premium-efficiency motor is $7000 and $8500, respectively, calculate the simple payback period for using the premium-efficiency motor rather than the originally selected motor for this application. Take the electricity tariff to be $0.15/kWh.

Solution

$$\text{Power}_{\text{IN}} = \frac{\text{power}_{\text{OUT}}}{\text{motor efficiency}}$$

Estimated power input to the standard-efficiency motor, $P_1 = 37/0.90 = 41.1$ kW

Estimated power input to the premium-efficiency motor, $P_2 = 37/0.94 = 39.4$ kW

$$\text{Savings in power } (P_1 - P_2) = (41.1 - 39.4) \text{ kW}$$

$$= 1.7 \text{ kW}$$

$$\text{Energy savings} = \text{power savings (kW)} \times \text{operating hours}$$

$$\text{Annual energy savings} = 1.7 \times 24 \times 365 \text{ kWh/year}$$

$$= 14,892 \text{ kWh/year}$$

$$\text{Annual cost savings} = \text{annual energy savings in kWh} \times \text{energy tariff}$$

$$= 14,892 \times 0.15$$

$$= \$2233.80/\text{year}$$

$$\text{Simple payback period} = \text{incremental cost of new motor/annual cost savings}$$

$$= (\$8500 - \$7000)/\$2233.80$$

$$= 0.7 \text{ years} \quad \blacktriangle$$

2.10   Input Power Supply to Motors

Since motor windings are inductive coils, there are two components of power. One is the actual power absorbed by the motor to do useful work called real power (or active power) and the other, the reactive power, is used for magnetizing the magnetic elements. The apparent power is the vector sum of the reactive and active power as shown in Fig. 2.22 and is normally computed in kVA (product of volts and amperes divided by 1000).

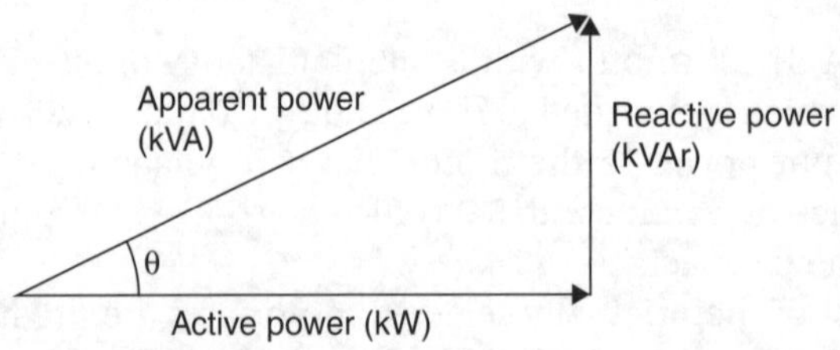

Figure 2.22   Vector diagram for power.

PF is the ratio of the active power to the apparent power. PF ranges from 0 to 1.0. The highest PF of 1.0 is achieved if there is no reactive power as in the case of totally resistive loads.

$$PF = \cos\theta = \frac{\text{active power (kW)}}{\text{apparent power (kVA)}}$$

2.10.1 Input Power

$$\text{Input power, kW} = \text{kVA} \times \text{PF}$$

$$\text{Input power, kW} = (\sqrt{3} \times V \times I \times PF)/1000$$

For a balanced load (Fig. 2.23),

$$V = \text{mean of line-to-line voltages, } (V_{ab} + V_{bc} + V_{ca})/3$$

$$I = \text{mean of line currents} = (I_a + I_b + I_c)/3$$

$$PF = \text{mean PF for the three phases}$$

Example 2.8   The following data have been measured for a motor using a power meter. Compute the input power to the motor.

$V_{ab} = 414\text{ V}$ $I_a = 50\text{ A}$ $PF_a = 0.87$

$V_{bc} = 415\text{ V}$ $I_b = 49\text{ A}$ $PF_b = 0.89$

$V_{ca} = 413\text{ V}$ $I_c = 51\text{ A}$ $PF_c = 0.88$

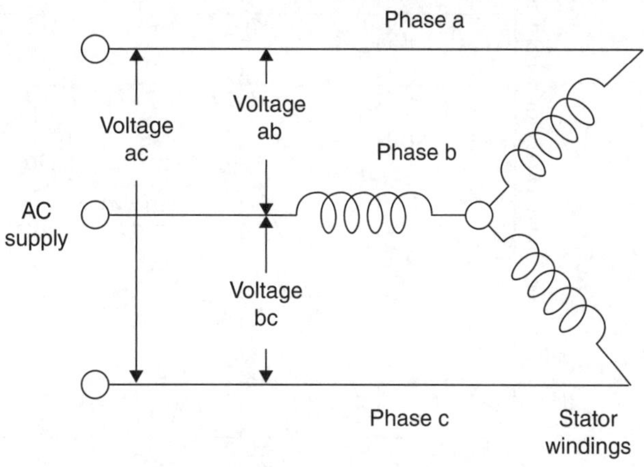

FIGURE 2.23   Power supply to a three-phase motor.

Solution

$$\text{Mean of line-to-line voltages} = (414 + 415 + 413)/3$$
$$= 414\,\text{V}$$
$$\text{Mean of line currents} = (50 + 49 + 51)/3$$
$$= 50\,\text{A}$$
$$\text{Mean PF} = (0.87 + 0.89 + 0.88)/3$$
$$= 0.88$$
$$\text{Input power to motor} = (\sqrt{3} \times 414 \times 50 \times 0.88)/1000$$
$$= 31.55\,\text{kW} \quad \blacktriangle$$

2.10.2   Supply Voltage

The performance of motors depends on the supply voltage. Therefore, motors are rated to operate at a particular supply voltage, and if the supply voltage deviates by more than ±10% from the rated voltage, it can significantly affect the performance of the motor.

If the supply voltage to the motor is below the rated value, a higher current will be drawn by the motor to produce the torque required by the load. Similarly, when the supply voltage increases, the magnetizing current increases, leading to saturation of the core iron, overheating, and higher current draw. Hence, both lower and higher supply voltages result in higher resistance (I^2R) losses for the stator and rotor of the motor and a drop in motor efficiency, as shown in Fig. 2.24. From the figure, it should also be noted that motor torque varies based on the square of the applied voltage. Therefore, a 10% drop in voltage results in a torque reduction of about 20%.

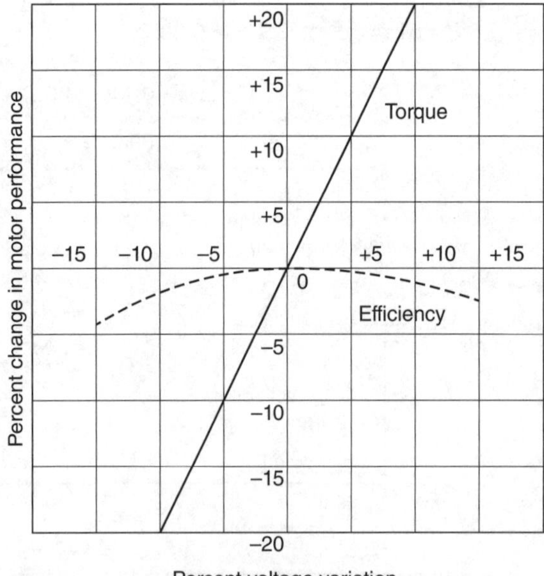

FIGURE 2.24   Variation in torque and motor efficiency.

In addition, unbalance in voltage between the different phases could also occur. Voltage unbalance at the motor windings causes the phase currents to be unbalanced, resulting in higher losses, overheating of the motor, and reduced life.

It is generally recommended for the voltage unbalance between the phases to be less than 1%. If the unbalance in voltage is more than 1%, derating of motors is required to avoid damage.

NEMA defines voltage unbalance as follows:

$$\text{Voltage deviation, }\% = \frac{\text{maximum voltage deviation from the mean of three phases}}{\text{average voltage}} \times 100$$

Example 2.9 Compute the voltage deviation based on the following measured data:

$$V_{ab} = 414\text{ V} \qquad V_{bc} = 415\text{ V} \qquad V_{ca} = 413\text{ V}$$

Solution

$$\text{Mean of three phases} = (414 + 415 + 413)/3$$

$$= 414\text{ V}$$

$$\text{Maximum deviation from the mean of three phases} = (415 - 414) = 1\text{ V}$$

$$\text{Voltage deviation, }\% = (1/414) \times 100$$

$$= 0.24 \quad \blacktriangle$$

2.10.3 Power Factor

As described earlier, PF is the ratio of the real power to the apparent power. Induction motors typically operate at a PF of about 0.85. A low PF leads to higher line losses, in addition to resulting in penalty charges from the utility companies. When the PF is low, higher line currents are required to meet the real power load, resulting in higher resistance (I^2R) losses.

Poorly loaded motors, due to oversizing or during idling, can also have a low PF. Since PF is the ratio of active power to apparent power, it can also be expressed as the ratio of active power to the vector sum of the active power and reactive power.

$$PF = \frac{\text{active power (kW)}}{\text{active power (kW)} + \text{reactive power (kVAr)}}$$

At low loading, the active power drops but the reactive power remains almost constant, resulting in a lower PF.

A low PF can be improved by installing capacitors in parallel to the load to reduce the reactive power. The leading capacitive current is used to offset the lagging inductive current, as shown in Fig. 2.25.

The PF correction can be "static correction" where capacitors are connected at each load (Fig. 2.26) or "bulk correction" where capacitors are connected at the distribution boards (Fig. 2.27).

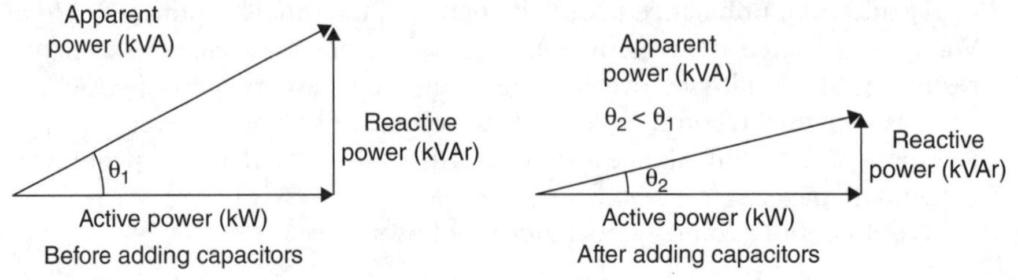

FIGURE 2.25 Vector diagrams showing improvement in PF.

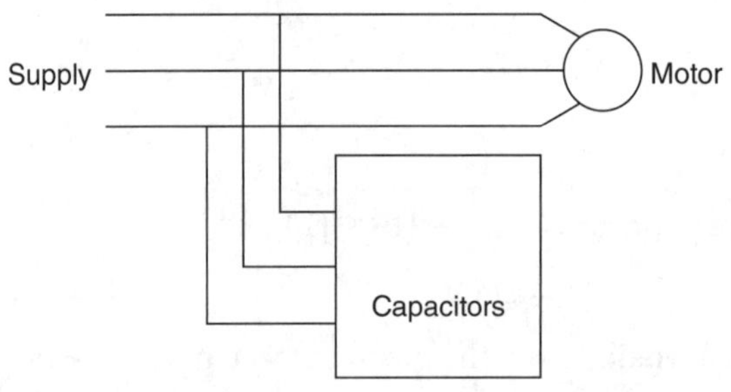

FIGURE 2.26 Static PF correction.

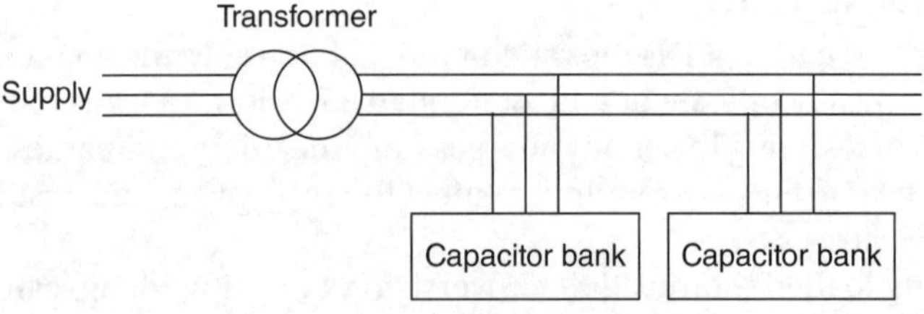

FIGURE 2.27 Bulk PF correction.

Example 2.10 Consider an industrial plant which is operating at an active power demand of 1500 kW and an apparent power demand of 1850 kVA. Estimate the present PF. Thereafter, calculate the kVAr rating of a suitable capacitive load that should be added to improve the PF to 0.9.

Solution Before PF correction (refer to Fig. 2.28):

$$\text{Active power} = 1500 \text{ kW}$$

$$\text{Apparent power} = 1850 \text{ kVA}$$

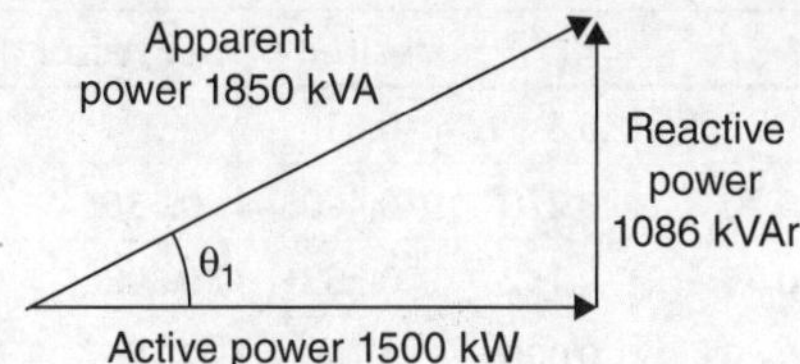

FIGURE 2.28   Vector diagram before PF correction.

$$PF = active\ power/apparent\ power$$

$$PF,\ \cos\theta_1 = 1500/1850 = 0.81$$

$$\theta_1 = 35.9°$$

$$\tan\theta_1 = reactive\ power/active\ power$$

$$\text{Therefore, reactive power} = \tan\theta_1 \times active\ power$$

$$= \tan(35.9°) \times 1500$$

$$= 1086\ kVAr$$

After PF correction (refer to Fig. 2.29):

$$PF = active\ power/apparent\ power$$

$$\text{Therefore, apparent power} = active\ power/PF$$

$$= 1500\ kW/0.9$$

$$= 1667\ kVA$$

$$\theta_2 = \cos^{-1}(0.9) = 25.8°$$

$$\text{Reactive power} = \tan(25.8°) \times 1500 = 725\ kVAr$$

$$\text{Therefore, size of capacitor required} = 1086 - 725 = 361\ kVAr\ \text{(capacitive)}\quad\blacktriangle$$

The capacitive load required for improving the PF can also be estimated using multiplier factors available in reference tables. An example of such data is shown in Table 2.7 where the intersection of the current PF and desired PF provides the multiplier factor to compute the capacitive load per kW.

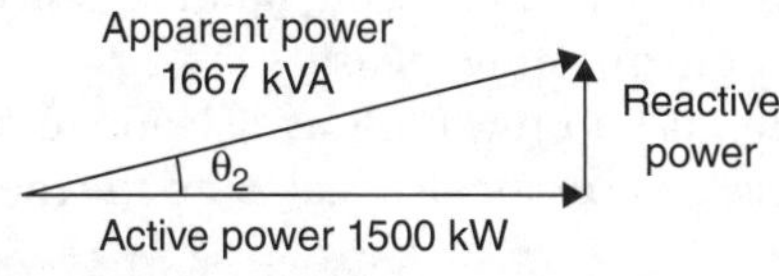

FIGURE 2.29   Vector diagram after PF correction.

	Desired power factor				
	0.8	0.85	0.9	0.95	1.0
0.70	0.270	0.400	0.536	0.692	1.02
0.75	0.132	0.262	0.398	0.553	0.882
0.80	0.000	0.130	0.266	0.421	0.750
0.85	0.000	0.000	0.135	0.291	0.620
0.90	0.000	0.000	0.000	0.156	0.484
0.95	0.000	0.000	0.000	0.000	0.329
1.0	0.000	0.000	0.000	0.000	0.000

(Row labels under "Original PF")

TABLE 2.7 Power Factor Correction Multiplier Values

Example 2.11 Compute the capacitive load required to improve the PF from 0.7 to 0.85 for an active power load of 500 kW.

Solution From Table 2.7, the intersection of the present PF of 0.7 and desired PF of 0.85 indicates that the multiplier is 0.4.

Therefore, the capacitive load is 500 kW × 0.4 = 200 KVAr. ▲

2.11 Maintenance of Motors

Proper maintenance is required to ensure reliability and good performance of motors. Motors can be allowed to operate until they break down but, this would result in disruptions to operations and high cost to repair them. To avoid breakdowns and unplanned stoppages, preventative maintenance programs are used for motors.

Preventive maintenance programs for motors involve tasks such as periodic lubrication of bearings, cleaning, replacement of rubber components of drive couplings, belts, and replacement of bearings.

In addition to preventive maintenance programs, predictive maintenance programs where periodically various measurements are taken and the data analyzed to predict any impending failure are also used for critical equipment. Typical data collected in a predictive maintenance program for motors include vibration levels, motor temperature, and resistance of the motor windings.

Some good maintenance practices for motors are listed below.

2.11.1 Lubrication

Generally, high-capacity motors require periodic greasing of bearings. However, since contaminated grease can lead to bearing failures, it is essential to ensure that the fittings are cleaned prior to injecting grease.

Overgreasing should be prevented as it leads to increased friction, resulting in bearing failure. Excess grease can also leak onto the motor windings, which can cause failure.

Care should be taken to ensure that the correct type of grease recommended by the manufacturer is used.

2.11.2 Cleaning

Regular cleaning of motors is essential as dirt can enter the motor and attack the insulation, as well as contaminate lubricants and damage bearings. Also, dirt buildup on the motor enclosure and fan cover can lead to the motor operating at a higher temperature, which will reduce the efficiency of the motor and its operating life.

If the airflow at the fan discharge is weak, it may indicate that the internal passages are blocked. The motor should then be taken out of service and cleaned.

Severe corrosion may indicate deterioration of internal surfaces. Such motors should also be checked, cleaned, and painted where necessary.

2.11.3 Motor Temperature

High motor temperature leads to failure and is an indication of other motor problems. Some of the common reasons for high motor temperature are a wrong selection of motors (undersizing of motor or wrong motor starting torque characteristics) and overloading of the motor. In addition, poor cooling due to dirt accumulation and damage to the cooling fan can also result in a high motor operating temperature.

Excessive heat can damage the insulation, which shortens the winding life (and therefore motor life). Therefore, the motor operating temperature should be checked periodically.

2.11.4 Voltage Testing

Supply voltage to motors should not deviate by more than $\pm10\%$ of the design nominal voltage. High deviations in voltage can lead to decreased motor efficiency and a shorter motor life. Similarly, unbalanced phase voltages can lead to high rotor currents, resulting in higher temperatures and high motor losses. Therefore, regular voltage measurements when the motor is operating can help identify problems and ensure efficient operation.

2.11.5 Insulation Testing

Periodic resistance testing of motor insulation in predictive maintenance programs can help identify degradation of insulation. The most common insulation test is a "megger test" which involves injecting a DC voltage of 500 to 1000 V to the motor and testing the resistance of the insulation. NEMA standards require a minimum resistance to ground of 1 MΩ/kV of rating plus 1 MΩ at 40°C ambient conditions. Motors in good condition will normally show readings of more than 50 MΩ. Low readings indicate a reduction in insulation resistance, which can be a result of moisture ingress, deterioration from age, or excessive heat.

Measured data can be trended to indicate a deterioration of motor windings. Where required, motors can be taken out of service and sent for cleaning and baking to avoid the need for a complete rewinding.

Such testing is specially required for motors that operate in wet, corrosive, and hot environments.

2.11.6 Vibration

Motor vibration can be caused by misalignment of the motor shaft, loose mountings, unbalanced rotor, or damaged bearings. Motor vibration sometimes can also be transmitted from the driven load. Vibration can cause damage to winding by loosening, cracking, or abrasion. Vibration can also lead to premature bearing failure.

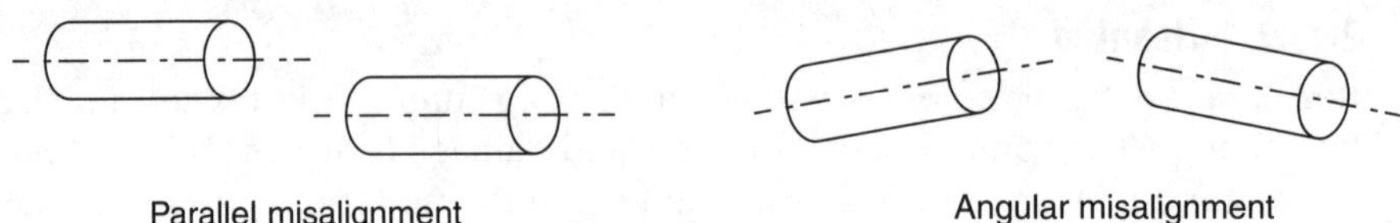

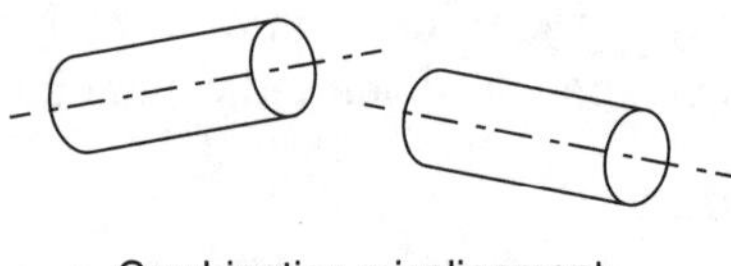

FIGURE 2.30   Types of shaft misalignment.

Therefore, motor vibration levels should be monitored periodically and trended to indicate potential problems so that they can be rectified before causing damage.

2.11.7 Alignment

It is important to ensure that the motor shaft and the shaft of the driven load are aligned as misalignment can lead to vibration, noise, damage to coupling, premature failure of bearings, as well as lower transmission efficiency.

Ideal alignment is achieved when the centerlines of the motor shaft and driven load are in line with each other. There are three basic types of misalignment, as shown in Fig. 2.30. Angular misalignment occurs when the motor shaft is at an angle to the shaft of the driven load, while parallel misalignment occurs when the centerlines of the two shafts are parallel to each other. Combination misalignment occurs when both angular and parallel misalignments are present.

2.12 Motor Drives

A typical motor-driven system is shown in Fig. 2.31 where a transmission drive system is normally installed between the motor and the driven load. The main reasons why a transmission drive system is used include the following:

- Aligning the motor and the load
- Increasing the torque
- Changing the speed of rotation
- Changing the direction of rotation
- Absorbing the impact of fluctuating or shock loads

The most commonly used types of transmission drive systems are as follows:

- Couplings
- Belt drives

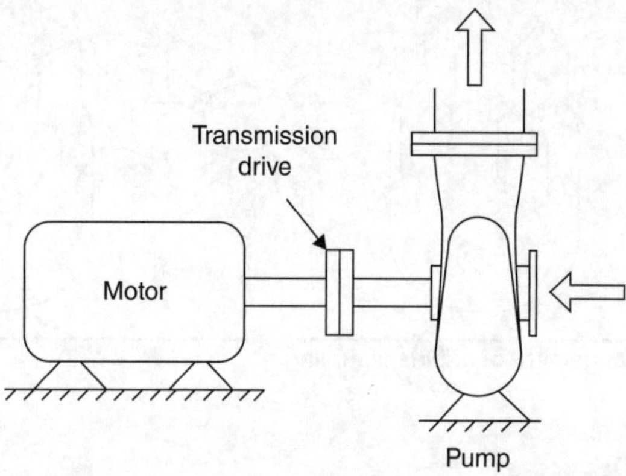

FIGURE 2.31 Arrangement of a typical motor-driven system.

- Gear drives
- Chain and sprockets

2.12.1 Couplings

In some applications, the motor torque is directly transmitted to the load through a coupling. Couplings can be rigid type or flexible type. Rigid couplings do not allow misalignment between the motor shaft and load shaft. Two common types of rigid couplings are sleeve type and flange type (Fig. 2.32). Sleeve-type couplings consist of a tube with an internal keyway to match those on the motor shaft and load shaft and tapped holes for locking of the sleeve. Flange couplings consist of two flanges mounted on the motor shaft and load shaft and connected together by a number of bolts.

Common types of flexible couplings used are pin, spider, and fluid couplings. Pin couplings are similar to the rigid flange couplings but use pins with rubber bushes, instead of bolts, to allow some degree of flexibility.

Spider couplings have two halves of coupling with a number of lugs on each, mounted on the motor shaft and load shaft (Fig. 2.33). The coupling halves are joined together with a rubber insert with "legs" to fit between the lugs on the two sides. This type of couplings also allows some misalignment and can absorb some of the vibration between the motor and the load.

Fluid couplings consist of a housing with two sets of vanes and a hydraulic fluid. When the motor shaft rotates, the set of drive vanes rotates and acts as a pump to pump the hydraulic fluid, which results in the rotation of the driven set of vanes connected to the load shaft. The highest efficiency possible for hydraulic couplings is about 94%.

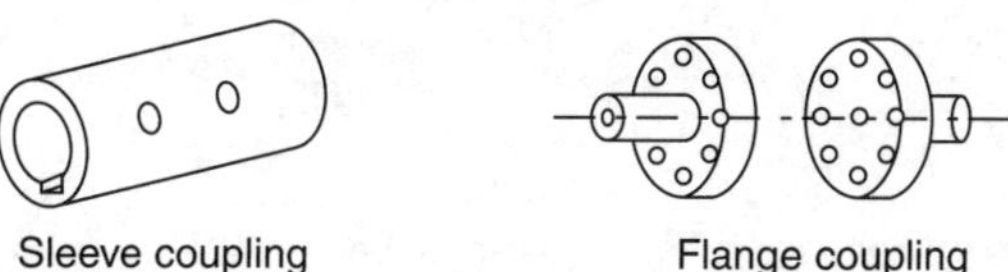

FIGURE 2.32 Rigid couplings.

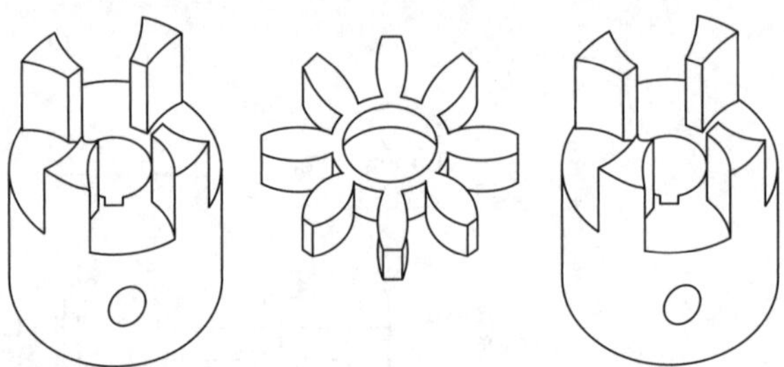

Figure 2.33   Arrangement of a spider coupling.

2.12.2   Belt Drives

Many applications of motors in buildings and industrial plants, such as fans, compressors, and conveying systems, use belt drives. They consist of pulleys on the shafts of the motor and load and one or more belts to convey the torque from the motor to the load. Since belt drives are relatively flexible, they are able to allow some misalignment between the two shafts. Belt drives also allow change in rotation speed between the motor and the load by changing the diameter of the two pulleys. Belt drives are relatively efficient and require little maintenance.

There are different types of belt drives and the most common types are discussed below.

V-belts

V-belts are trapezoidal in cross-section (Fig. 2.34) and fit into grooves on the respective pulleys. Usually, multiple belts are used depending on the load transmission requirements. The efficiency of V-belt drives depends on factors such as design, pulley sizing, and torque, but generally have a high efficiency ranging from 95% to 98%. However, the efficiency drops significantly when belt slippage occurs due to belt wear. Therefore, periodically, the belt drives need to be checked and retensioned increasing the distance between the two pulleys or replacing the belts.

Cogged Belts

Cogged belts are similar to V-belts but have additional slots perpendicular to the belt to reduce the bending resistance of the belt. The transmission efficiency of cogged belts is about 2% higher than that of standard V-belts.

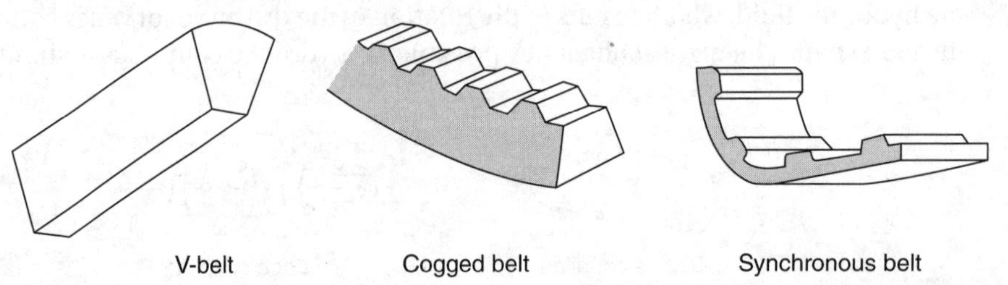

Figure 2.34   Common types of belts.

Synchronous Belts

Synchronous belts, which are also called timing belts, are "toothed" (Fig. 2.34) and fit into matching toothed-drive sprockets. Synchronous belts operate at an efficiency of about 98% and are able to maintain the efficiency at high torque conditions as their design prevents belt slippage (unlike in V-belt drives). Synchronous belts can operate in wet and oily conditions without slipping but cannot be used for shock loads.

2.12.3   Gear Drives

Gear drives consist of two or more intermeshing gear wheels between the drive system and the load. The main applications of gear drives include changing speed, increasing torque, and changing the direction of rotation.

Some of the common types of gears are as follows:

- Spur gears
- Helical gears
- Bevel gears
- Worm gears

Spur gear drives consist of gear wheels with straight teeth parallel to the axis of rotation, which mesh together to transmit torque (Fig. 2.35). They are able to change the speed, torque, and direction of rotation.

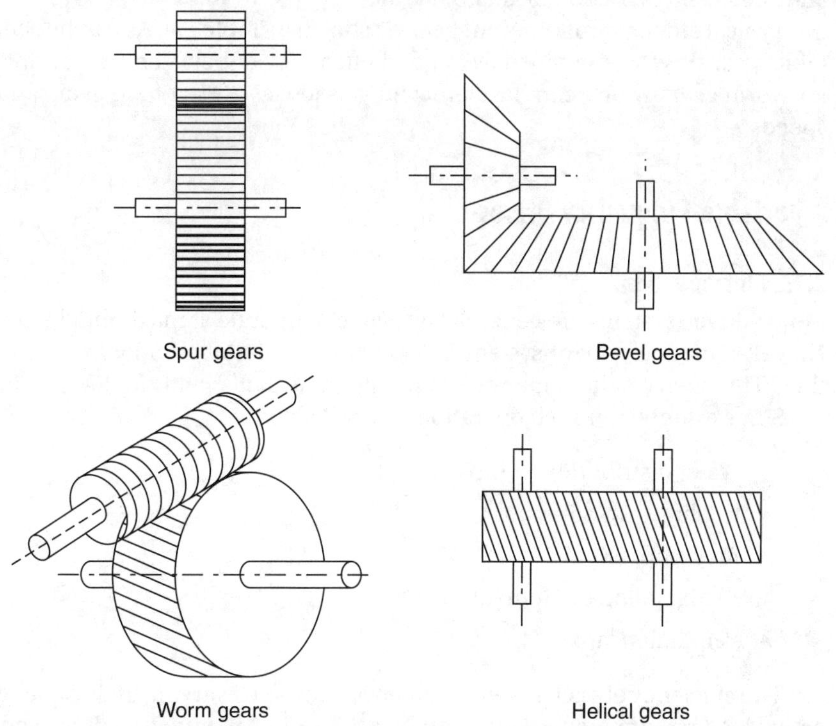

Figure 2.35   Common types of gears.

Type of gear	Typical efficiency (%)
Spur	98–99
Helical	98–99
Bevel	98–99
Worm	20–98

TABLE 2.8 Efficiency of Gears

Helical gears are similar to spur gears but with angled teeth to enable gradual engagement of the teeth during rotation, which results in smoother operation.

Bevel gears consist of two gear wheels shaped like cones but with the tips removed. They are mounted on shafts which can be at different angles relative to each other (except 0° and 180°), depending on the number of teeth on the two gears.

Worm gears consist of a drive worm (screw) and a driven wheel (Fig. 2.35). Worm gears can reduce speed, transmit high torque, and change the direction of rotation between the drive and driven shafts by 90°. Worm gears are self-locking as the rotational transmission can only take place in one direction because the wheel cannot turn the worm.

Losses in gear drives arise mainly due to friction losses that take place when the gears mesh. Friction losses depend on factors such as tooth profile, speed reduction, and coefficient of friction of gear material.

Typical efficiency of different gears is shown in Table 2.8. As can be seen from the table, gear drives generally have a good efficiency of between 98% and 99%, except for worm gears which can have efficiency as low as 20% at high gear ratios and low speeds.

2.13 Variable-Frequency Drives

2.13.1 Introduction

Motor-driven systems are generally designed to operate at maximum load conditions. However, the majority of systems operate at their full load only for short periods of time. This often results in many systems operating inefficiently for long periods of time.

Some examples of such operations are as follows:

- Water distribution pumps
- Cooling tower fans
- Air compressors
- Refrigeration equipment
- Ventilation fans

The efficiency of such systems can be improved by varying their capacity to match actual load requirements, which can be achieved by modulating the speed of motors using variable-frequency drives (VFDs). A typical VFD is shown in Fig. 2.36.

FIGURE 2.36 A typical VFD unit (Danfoss VLT® HVAC Drive, courtesy of Danfoss).

2.13.2 Components of a VFD

As explained earlier in Sec. 2.2, the speed of motors depends on the number of poles and frequency of the power supply where

$$\text{Synchronous speed (rpm)} = (120 \times \text{power supply frequency in Hz})/(\text{number of poles in the motor})$$

Therefore, for a motor with a fixed number of poles, the speed can be varied by changing the frequency of the power supply.

A VFD (sometimes called VSD or variable speed drive) is an electronic unit which provides infinitely variable control of the speed of three-phase AC motors by converting the fixed mains voltage and frequency into variable quantities (Fig. 2.37). It has no moving parts and uses a rectifier which is connected to the mains supply to generate a

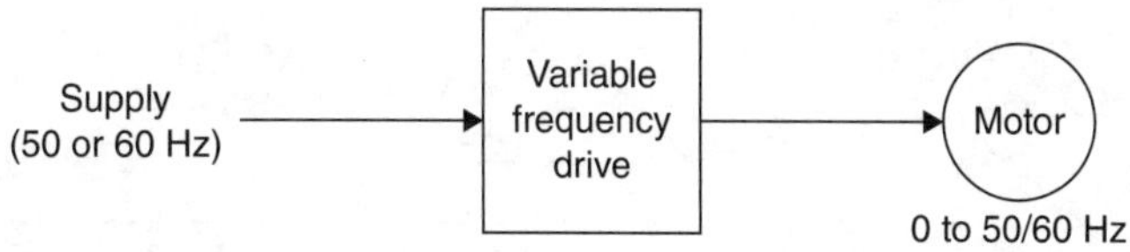

FIGURE 2.37 Operation of a VFD unit.

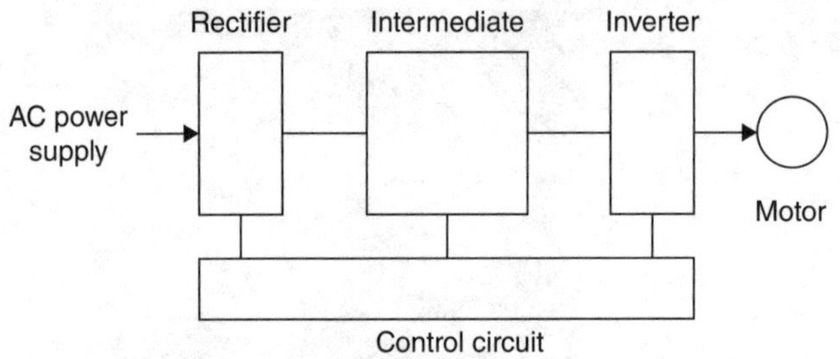

Figure 2.38   Configuration of a typical VFD.

pulsating DC voltage and direct current, which is then passed to an inverter to generate the frequency of the motor voltage.

The two main techniques used for generating the required output frequency are pulse amplitude modulation (PAM) and pulse width modulation (PWM). In PAM, the amplitude of the voltage pulse is varied, while in PWM, the width of the voltage pulse is varied to generate the required frequency. PWM inverters are generally considered to be more efficient than PAM type.

The configuration of a typical VFD is shown in Fig. 2.38. The main components of a VFD are the rectifier, inverter, intermediate circuit, and control circuit.

The main functions of each component are listed below.

Rectifier

It converts the AC to DC. The arrangement of a three-phase rectifier is shown in Fig. 2.39.

Intermediate Circuit

The intermediate circuit, also called DC bus or DC link, stabilizes or smoothens the pulsating DC voltage and reduces the feedback of harmonics to the mains supply.

The circuit mainly consists of an inductor and capacitor (Fig. 2.40) to filter any AC component from the DC waveform.

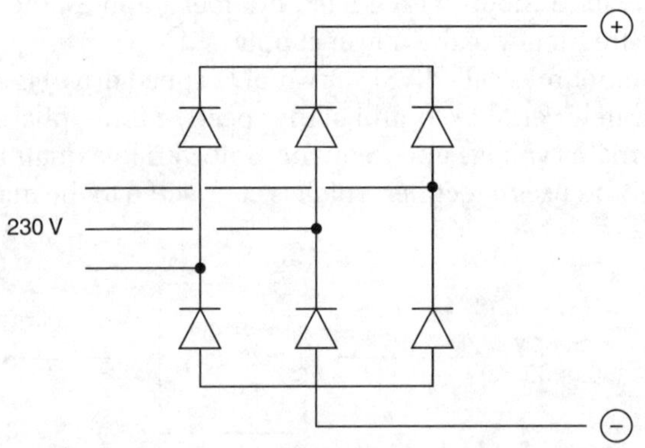

Figure 2.39   Arrangement of a three-phase rectifier.

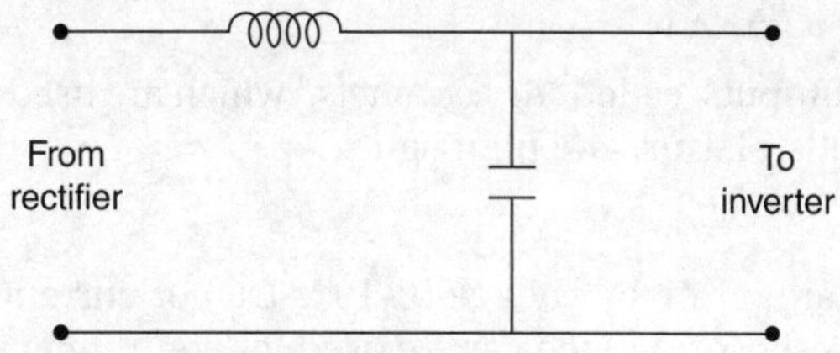

FIGURE 2.40   Arrangement of an intermediate circuit.

Inverter

The inverter converts the filtered DC voltage into a pulsating DC waveform. By varying the output of the inverter, the pulsating DC waveform is used to simulate an AC waveform with variable voltage and frequency.

The inverter normally uses insulated-gate bipolar transistors (IGBTs) to switch on and off the DC voltage.

Control Circuit

It controls the operation of the VFD, including firing of the rectifiers and transistors.

A simplified arrangement of a three-phase VFD showing the interconnection between the different components is shown in Fig. 2.41.

2.13.3   Features of VFDs

Today's VFDs come with many built-in functions and features which make them go beyond being devices to vary the voltage and frequency of the power supply to motors. VFDs are able to take different forms of inputs, which can be processed by the built-in controller, while various outputs can also be provided for monitoring by external systems.

The main types of inputs and outputs are listed below.

Digital Inputs

Digital inputs are used to interface the VFD with start/stop push button switches, selector switches, and relay contacts from various control circuits.

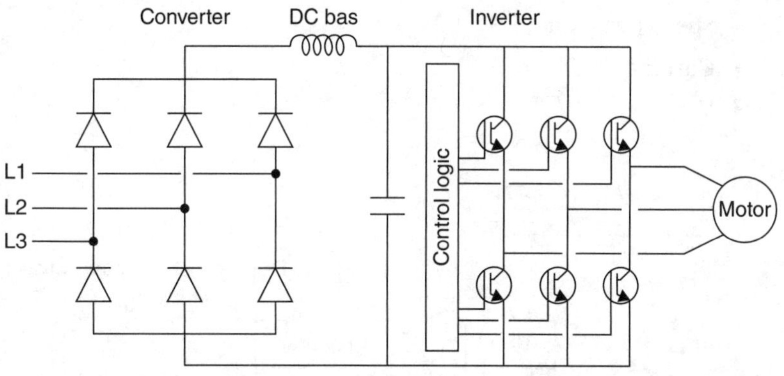

FIGURE 2.41   Simplified arrangement of a three-phase VFD.

Digital Outputs

They are relay outputs, called "dry contacts," which are used to switch external devices such as solenoids, alarms, and pilot lights.

Analog Inputs

Analog inputs are normally voltage (0–10 V DC) or current (4–20 mA) from external sources such as sensors, which can be used for controlling the output from the drive. Most VFDs can also take direct input from sensors such as RTDs (resistance temperature detectors).

Analog Outputs

Analog outputs are provided to external devices to monitor parameters such as the current and speed of the drive.

Figure 2.42 shows a typical speed control system for a pump using the line pressure as the input. The pressure sensor provides an analog input such as 0 to 10 V DC or 4 to 20 mA from its transmitter to the VFD controller. The controller uses this input to compare with the required set-point to vary the speed of the pump.

Adjustable Parameters

Various parameters used to operate a VFD are user-adjustable so that they can be set to meet the specific needs of each system. Some common parameters that can be user defined are as follows:

- Minimum and maximum speeds
- Control set-points
- Ramp-up and ramp-down rates
- PID (proportional, integral, and derivative) values for control

Performance Parameters

The performance of most VFDs can also be monitored during operation. The monitoring can be either local, using the display panel on the VFD, or external, using the communication port of the drive. Some typical performance parameters that can be monitored are as given below:

- Speed (frequency)
- Current
- Voltage

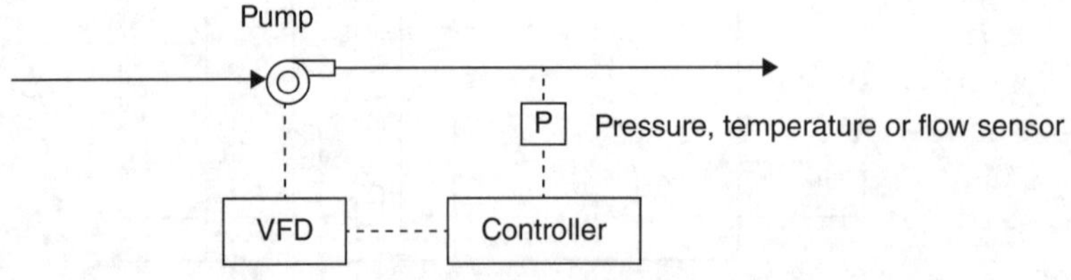

FIGURE 2.42 Typical arrangement of a VFD used for pump speed control.

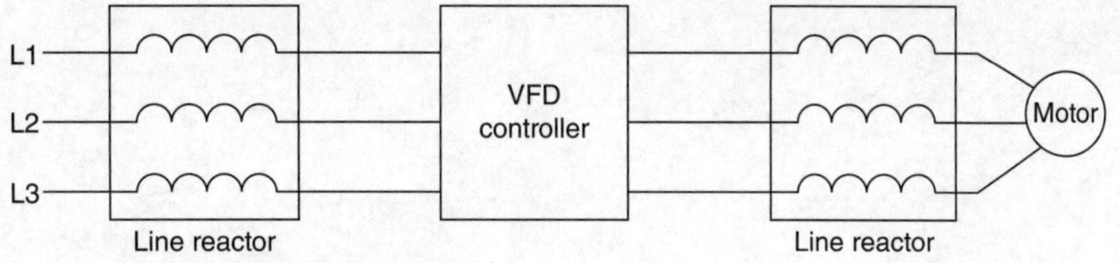

FIGURE 2.43   Arrangement of line reactors.

- Power factor
- Power (kW)
- Cumulative energy consumption (kWh)
- Cumulative running hours

2.13.4 Selection and Installation

Type of Drive

Right selection and installation is important for ensuring problem-free operation of VFDs. The drive unit should be selected based on the load characteristics. For most loads such as pumps and fans, the load torque reduces when speed reduces and, therefore, variable-torque drives can be used. However, for applications such as conveyors and traction drives, torque remains constant, irrespective of speed, and, hence, constant-torque drives are required.

Location

The installation location is also important to ensure the reliability and performance of VFDs. Drive units generate heat during operation and therefore need to be installed in a location where adequate ventilation is available. In addition, the location should be dry and relatively clean (free of dust and corrosive elements).

Another important aspect of VFD installation is the distance between the drive unit and the motor. Maximum allowable cable length between the drive and motor is usually specified by the drive manufacturer and should not be exceeded to prevent high-voltage spikes which can damage the motor.

Line Reactors

Line reactors, which are essentially inductors, are often installed before or after the drive to stabilize the current waveform and reduce harmonic distortion (Fig. 2.43). When installed before the drive, they help to stabilize the waveform of the current supplied to the drive, which may be distorted by nonlinear loads. Since VFDs generate harmonics when converting AC to DC and DC back to AC, line reactors installed between the drive and motor help smoothen the waveform by absorbing line spikes and filling voltage sags.

CHAPTER 3

Pumping Systems

Pumps are widely used in industrial plants and account for a significant portion of their energy consumption. They are used in cooling and heating systems as well as for transferring liquids in various process systems.

3.1 Types of Pumps

The two main types of pumps are centrifugal and positive displacement. Centrifugal pumps are the most common type of pumps used. They have an impeller mounted on a shaft, which is driven by a motor and rotates in a volute or diffuser casing. The rotating impeller imparts kinetic energy to the fluid, which is later converted to a static head in the volute casing. Centrifugal pumps can be further classified as radial flow (liquid enters through the center of the impeller and is pushed out along the impeller blades at right angles to the shaft), axial flow (liquid is pushed parallel to the shaft like a propeller), and mixed flow (combination of radial and axial flow).

Positive-displacement pumps operate by displacing a fluid from a fixed volume due to the action of a piston stroke or shaft rotation. Types of positive-displacement pumps include piston, plunger, diaphragm, gear, lobe, screw, and sliding vane. Classification of common types of pumps is shown in Fig. 3.1.

Centrifugal pumps are more commonly used as they are easier to operate and encounter less wear. On the other hand, positive-displacement pumps suffer more wear and also require relief valves as a safety precaution to prevent overpressurizing the system in a situation when all the downstream valves are closed.

Therefore, positive-displacement pumps are used only for special applications such as pumping highly viscous fluids, when the flow needs to be metered, and high-pressure low-flow applications, when the flow needs to remain constant with variation in the pump head.

3.2 System and Pump Curves

Pumping systems are either closed, where a fluid is circulated in a closed loop, or open, where static pressure is present due to a height difference such as with cooling towers or open vessels. Arrangements of closed and open systems are shown in Fig. 3.2.

Pumps and pumping systems are normally rated based on the pressure head and flow rate. These two parameters are dependent on each other.

The pressure developed by a pump is necessary to overcome resistances in the system such as those due to frictional losses in piping; pressure losses across valves, coils,

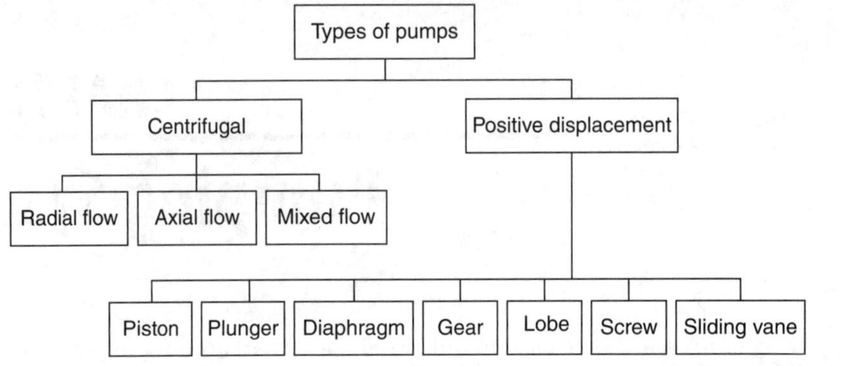

FIGURE 3.1   Classification of pumps.

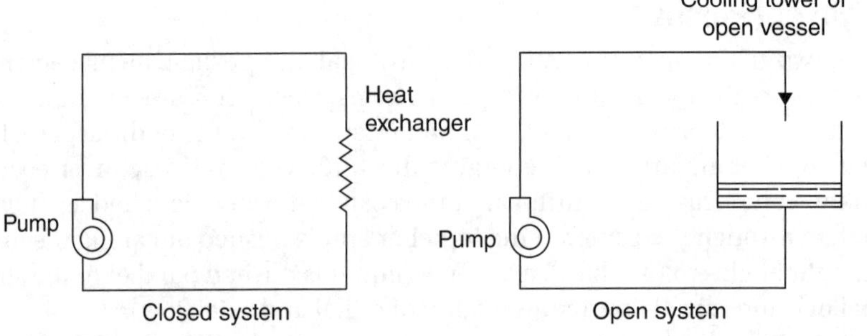

FIGURE 3.2   Arrangement of closed and open systems.

and heat exchangers; and static head differences in open systems. The relationship between head losses in a system and the system flow rate is called the *system resistance curve*. Typical resistance curves for open and closed systems are shown in Fig. 3.3.

The difference between the two curves is that for open systems the static pressure difference or independent pressure due to height difference is added to the system

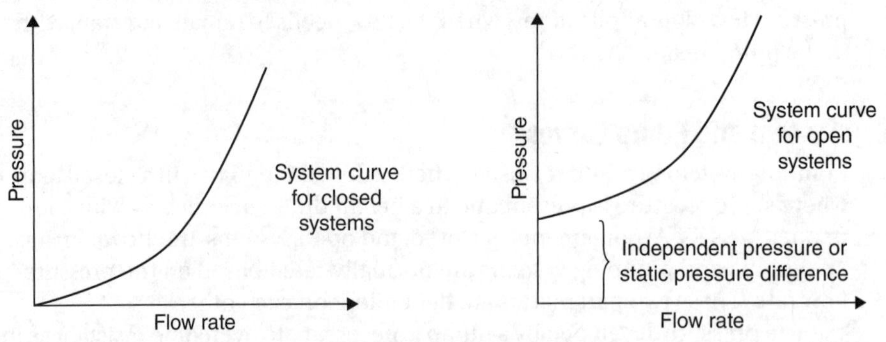

FIGURE 3.3   System resistance curves (closed and open systems).

curve. The system resistance curve is parabolic in shape since the pressure losses in the system are proportional to the square of the flow ($\Delta P \propto \text{flow}^2$).

Each particular system has its own system resistance curve due to its own unique pipe sizing, pipe lengths, and fittings. The system resistance curve, therefore, can change if components in the system are changed.

Example 3.1   A closed water-pumping system consists of one big pump and one small pump. When the big pump is in operation, the pressure differential across the pump is 3 bar and the water flow rate is measured to be 200 L/s. Estimate the water flow rate for the same system when the small pump is in operation and the pressure differential across the pump is 2 bar.

Solution   The pressure differential in the system is proportional to the square of the flow ($\Delta P \propto \text{flow}^2$).

$$\text{Pressure differential across the big pump } \Delta P_1 = 3 \text{ bar}$$

$$\text{Pressure differential across the small pump } \Delta P_2 = 2 \text{ bar}$$

$$\text{Water flow rate with the big pump } Q_1 = 200 \text{ L/s}$$

$$\text{Water flow rate with the small pump } Q_2 = Q_1 \times \sqrt{(\Delta P_2 / \Delta P_1)}$$

$$= 200 \times \sqrt{(2/3)}$$

$$= 163.3 \text{ L/s (see Fig. 3.4)} \quad \blacktriangle$$

Similarly, the relationship between the flow rate and pressure developed by a pump is called a *pump curve*. For centrifugal pumps, the flow rate is reduced when the discharge pressure is increased. For positive-displacement pumps, the flow rate is dependent on the operating speed of the pump and remains almost constant irrespective

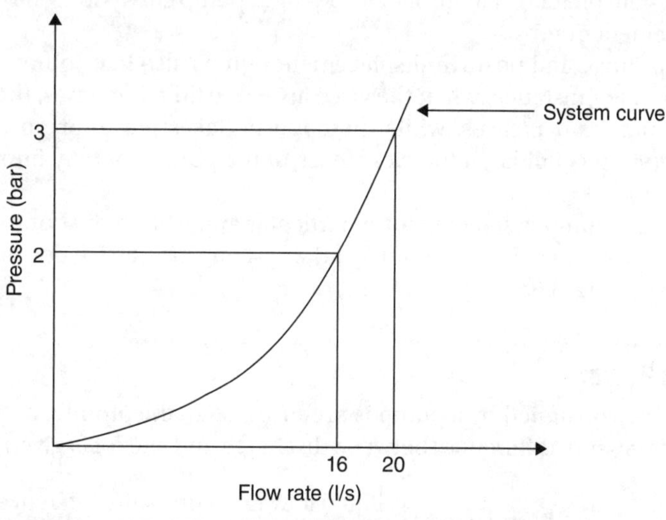

Figure 3.4   Figure for Example 3.1.

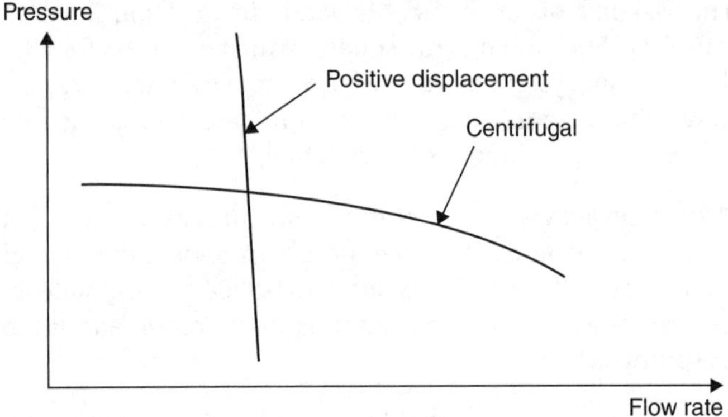

Figure 3.5   Typical pump curves.

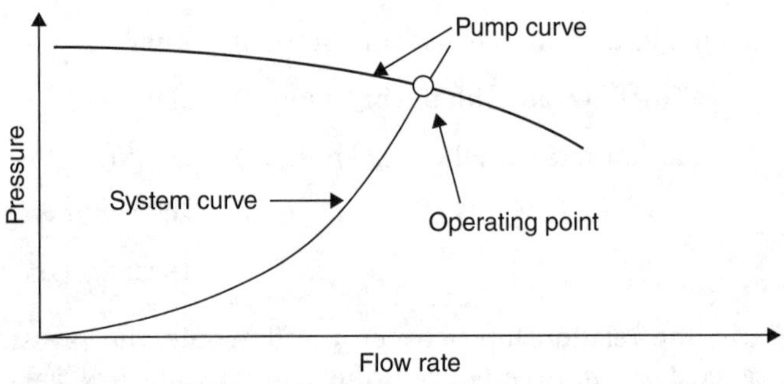

Figure 3.6   System and pump curves for a pumping system.

of the system pressure. Figure 3.5 shows typical pump curves for centrifugal and positive-displacement pumps.

Centrifugal and positive-displacement pumps also have other differences in performance. For instance, when the viscosity of the fluid increases, the flow rate for centrifugal pumps decreases, while the opposite takes place for positive-displacement pumps (viscous fluids fill the clearances of the pump, thereby improving volumetric efficiency).

When a pump is selected for a particular application, it should be one that has a performance curve that can intersect the system curve at the desired operating point, as shown in Fig. 3.6.

3.3   Pump Power

The power consumed by a pump is proportional to the product of the liquid flow rate and the pressure difference (between discharge and suction), given by

$$\text{Pump power} \propto \frac{\text{liquid flow rate} \times \text{pressure difference}}{\text{efficiency}} \tag{3.1}$$

In SI units:

$$\text{Pump impeller power (kW)} = \frac{\text{flow rate (m}^3/\text{s)} \times \text{pressure difference (N/m}^2)}{1000 \times \text{efficiency}} \quad (3.2)$$

In Imperial units:

$$\text{Pump brake horsepower} = \frac{\text{flow rate (USGPM)} \times \text{pressure difference (ft. water)}}{3960 \times \text{efficiency}} \quad (3.3)$$

Therefore, based on relationships shown in Eqs. (3.1) to (3.3), pump power consumption can be lowered by reducing the liquid flow rate and the pressure head and increasing the pump efficiency. These are the main strategies employed to reduce energy consumption of pumps and pumping systems.

3.4 Affinity Laws

The performance levels of centrifugal pumps under different conditions are related by the pump affinity laws given in Table 3.1. The pump affinity laws relate the pump speed and impeller diameter to the flow rate, pressure developed across the pump, and brake horsepower of the pump.

Example 3.2 A pump delivers 100 L/s of water and operates at 1400 rpm (revolutions per minute). The pump shaft power is 50 kW. If the pump speed is reduced to 1300 rpm using a variable speed drive, estimate the new pump flow rate and shaft power. Assume the pump efficiency remains the same.

Solution

$$Q_1 = 100 \text{ L/s}$$
$$P_1 = 50 \text{ kW}$$
$$N_1 = 1400 \text{ rpm}$$
$$N_2 = 1300 \text{ rpm}$$
$$Q_2 = Q_1 \times (N_2/N_1) = 100 \times (1300/1400) = 92.9 \text{ L/s}$$
$$P_2 = P_1 \times (N_2/N_1)^3 = 50 \times (1300/1400)^3 = 40 \text{ kW} \quad \blacktriangle$$

	Change in speed (N)	Change in impeller diameter (D)
Flow rate (Q)	$Q_2 = Q_1 \left(\dfrac{N_2}{N_1} \right)$	$Q_2 = Q_1 \left(\dfrac{D_2}{D_1} \right)$
Pressure difference (Δp)	$\Delta p_2 = \Delta p_1 \left(\dfrac{N_2}{N_1} \right)^2$	$\Delta p_2 = \Delta p_1 \left(\dfrac{D_2}{D_1} \right)^2$
Power (P)	$P_2 = P_1 \left(\dfrac{N_2}{N_1} \right)^3$	$P_2 = P_1 \left(\dfrac{D_2}{D_1} \right)^3$

TABLE 3.1 Pump Affinity Laws

3.5   Pressure Losses in Pipes and Fittings

When a liquid flows through a piping system, head or pressure losses take place due to fluid friction in the piping and resistance offered to the flow by the various devices such as valves, strainers, and bends installed in the piping system.

The friction losses depend on the pipe material, length of the piping, fluid velocity, and properties of the fluid. Therefore, for a given fluid such as water, the friction losses can be reduced by minimizing the pipe length and reducing the flow velocity (increasing pipe diameter).

On the other hand, for a particular flow velocity, losses due to various fittings and devices installed on piping systems depend on the design of the device or fitting. Therefore, for a given flow velocity, different types of valves can have different losses associated with them.

Pressure drop due to the friction of a fluid flowing in a pipe is given by the Darcy–Weisbach equation:

$$\Delta h = f \frac{L}{D} \frac{V^2}{2g} \tag{3.4}$$

where Δh = friction loss (Pa/m), f = friction factor (dimensionless), L = length of pipe (m), D = inside diameter of pipe (m), V = average velocity of fluid (m/s), and g = acceleration due to gravity (9.8 m/s^2).

As a result of the above relationship, the system pressure is proportional to the second power of the flow rate.

To simplify the design of piping systems, friction losses per unit length for various pipe sizes and materials are generally expressed in the form of charts and tables and are available in reference books.

In good piping designs, pipe sizes are usually selected to maintain friction losses to be equal to or less than 150 Pa/m. Table 3.2 provides a list of sizes for schedule 40 steel pipes selected to maintain the friction losses to be within 150 Pa/m for water pumping applications.

Water flow rate (L/s)	Recommended pipe diameter for schedule 40 steel pipes (mm)
5	50 or 65
10	100
20	125 or 150
40	200
60	200 or 250
80	250
100	250
200	350
300	400
400	450
600	500

TABLE 3.2   Size of Steel Pipes for Various Water Flow Rates

Nominal pipe diameter (mm)	90° regular elbow	90° long-radius elbow	45° long-radius elbow
25	0.43	0.41	0.22
40	0.4	0.35	0.21
50	0.38	0.3	0.2
100	0.31	0.22	0.18
150	0.29	0.18	0.17
200	0.27	0.16	0.17
250	0.25	0.14	0.16
300	0.24	0.13	0.16

Regular and long-radius elbows refer to a bend radius of 1 time and 1.5 times the pipe diameter, respectively.

TABLE 3.3(a)   Dynamic Loss Coefficients for Bends

Furthermore, pressure losses in fittings are also proportional to the square of the flow velocity and can be expressed as

$$\Delta h = K \frac{V^2}{2g} \tag{3.5}$$

where K = dynamic loss coefficient depending on type of fitting, size, and flow velocity.

The value of K is measured experimentally and is tabulated in reference books. Selected values of the dynamic loss coefficient are listed in Tables 3.3(a), (b), and (c).

Example 3.3   A 250-mm-diameter piping system used for pumping water at a rate of 100 L/s has ten 90° regular elbows. If the pump has a rated efficiency of 75% at the operating point, compute the reduction in annual pump motor energy consumption that would result if 45° long-radius elbows are used instead of the regular elbows. Assume motor efficiency to be 88% and the pump operating hours to be 24 hours a day.

Nominal pipe diameter (mm)	Inline Tee	Branch Tee
25	0.26	1.0
40	0.23	0.9
50	0.2	0.84
100	0.15	0.7
150	0.12	0.62
200	0.1	0.58
250	0.09	0.53
300	0.08	0.5

TABLE 3.3(b)   Dynamic Loss Coefficient for Pipe Tees

Nominal pipe diameter (mm)	Globe valve	Gate valve	Check valve
25	13	–	2
40	10	–	2
50	9	0.34	2
100	6.5	0.16	2
150	6	0.1	2
200	5.7	0.08	2
250	5.7	0.06	2
300	5.7	0.05	2

TABLE 3.3(c)   Dynamic Loss Coefficient for Selected Valves (Fully Open)

Solution

$$\text{Flow area} = \pi \times (\text{pipe diameter})^2/4$$

$$= 3.14 \times (0.25)^2/4$$

$$= 0.049 \text{ m}^2$$

$$\text{Flow velocity } v = \text{volume flow rate/flow area}$$

$$= 0.1 \text{ m}^3/\text{s}/0.049 \text{ m}^2$$

$$= 2 \text{ m/s}$$

Using Eq. (3.5):

$$\text{Losses across regular } 90° \text{ elbows} = \text{number of bends} \times K \text{ value for bends} \times v^2/(2 \times g)$$

$$= 10 \times 0.25 \text{ (Table 3.3a)} \times 2^2/(2 \times 9.81)$$

$$= 0.509 \text{ m (of water)}$$

$$\text{Losses across } 45° \text{ long-radius elbows} = \text{number of bends} \times K \text{ value for bends} \times v^2/(2 \times g)$$

$$= 10 \times 0.16 \text{ (Table 3.3a)} \times 2^2/(2 \times 9.81)$$

$$= 0.326 \text{ m (of water)}$$

Using Eq. (3.2):

$$\text{Reduction in pump power} = \frac{0.1 \times (0.509 - 0.326) \times 9.81 \times 1000 \ (\text{N/m}^2/\text{m})}{1000 \times 0.75 \times 0.88}$$

$$= 0.27 \text{ kW}$$

$$\text{Reduction in annual energy consumption} = 0.27 \ (\text{kW}) \times 24 \text{ h} \times 365 \text{ d}$$

$$= 2365 \text{ kWh} \quad \blacktriangle$$

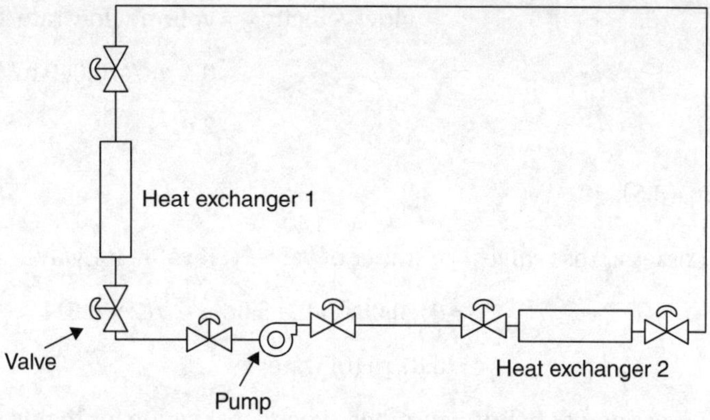

Figure 3.7   Diagram for Example 3.4.

Example 3.4   In the pumping system shown in Fig. 3.7, a pump circulates 100 L/s of water through two heat exchangers. Based on the following system information, estimate the total head for the pump in meters of water.

$$\text{Total pipe length} = 50 \text{ m}$$

$$\text{Pipe pressure loss} = 150 \text{ Pa/m}$$

$$\text{Diameter of pipes and fittings} = 250 \text{ mm}$$

$$\text{Number of pipe bends} = 15 \ (90° \text{ regular elbows})$$

$$\text{Number of gate valves} = 6$$

$$\text{Pressure loss across heat exchanger } 1 = 5 \text{ m of water}$$

$$\text{Pressure loss across heat exchanger } 2 = 3.5 \text{ m of water}$$

Solution

$$\text{Total pressure losses in piping} = \text{pipe length} \times \text{loss per unit length}$$

$$= 50 \text{ m} \times 150 \text{ Pa/m}$$

$$= 7500 \text{ Pa}$$

$$= 0.76 \text{ m (of water)}$$

$$\text{Total pressure losses across heat exchangers} = 5 \text{ m} + 3.5 \text{ m}$$

$$= 8.5 \text{ m (of water)}$$

$$\text{Water flow rate} = 100 \text{ L/s} = 0.1 \text{ m}^3/\text{s}$$

$$\text{Flow area} = \pi \times (\text{pipe diameter})^2/4$$

$$= 3.14 \times (0.25)^2/4$$

$$= 0.049 \text{ m}^2$$

$$\text{Flow velocity v} = \text{volume flow rate/flow area}$$

$$= 0.1 \ \text{m}^3/\text{s}/0.049 \ \text{m}^2$$

$$= 2 \ \text{m/s}$$

Using Eq. (3.5):

$$\text{Losses across valves} = \text{number of valves} \times \text{K value for valves} \times v^2/(2 \times g)$$

$$= 6 \times 0.06 \ (\text{Table 3.3c}) \times 2^2/(2 \times 9.81)$$

$$= 0.07 \ \text{m (of water)}$$

$$\text{Losses across bends} = \text{number of bends} \times \text{K value for bends} \times v^2/(2 \times g)$$

$$= 15 \times 0.25 \ (\text{Table 3.3a}) \times 2^2/(2 \times 9.81)$$

$$= 0.76 \ \text{m (of water)}$$

$$\text{Total head for pump} = \text{pipe losses} + \text{heat exchanger losses} +$$
$$\text{valve losses} + \text{bend losses}$$

$$= (0.76 + 8.5 + 0.07 + 0.76) \ \text{m}$$

$$= 10.09 \ \text{m (of water)} \quad \blacktriangle$$

3.6   Parallel and Series Pumping

Multiple pumps are used in many pumping applications. Pumps installed in parallel is the most common type of arrangement where two or more pumps of the same or different capacity operate in parallel, as shown in Fig. 3.8.

One of the main advantages of parallel pumping systems is the flexibility they offer to operate multiple pumps to best match varying load requirements. In such a pumping system, multiple pumps can be switched on or off depending on the load so that the operating pumps can operate more efficiently (Fig. 3.9).

In systems where high pumping pressures are required, pumps arranged in series, as shown in Fig. 3.10, are used. In series pumping arrangements, the multiple pumps need to be selected to have the same flow capacity (Fig. 3.11).

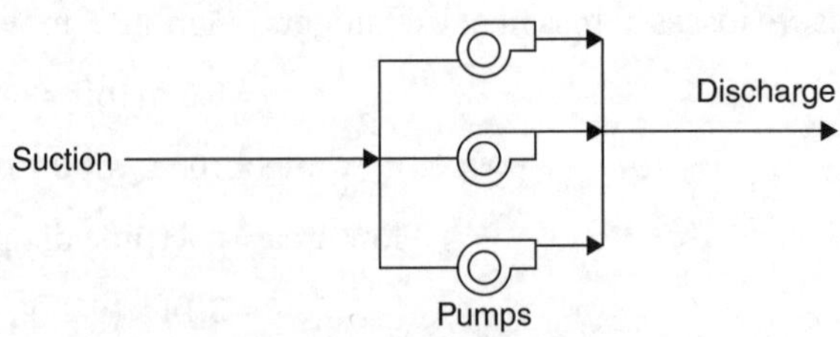

FIGURE 3.8   Arrangement of pumps operating in parallel.

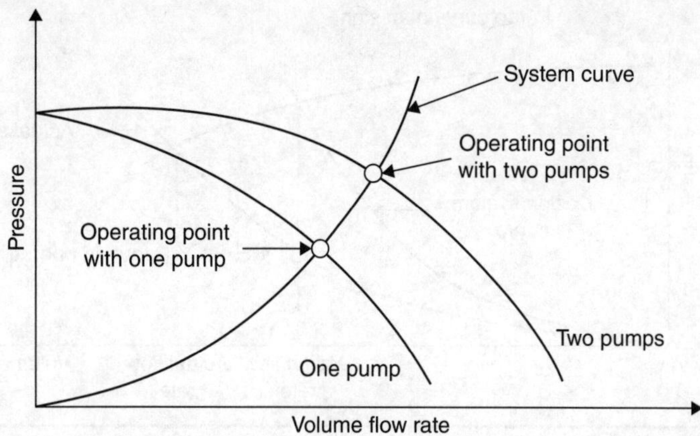

FIGURE 3.9   System curve for parallel pumps.

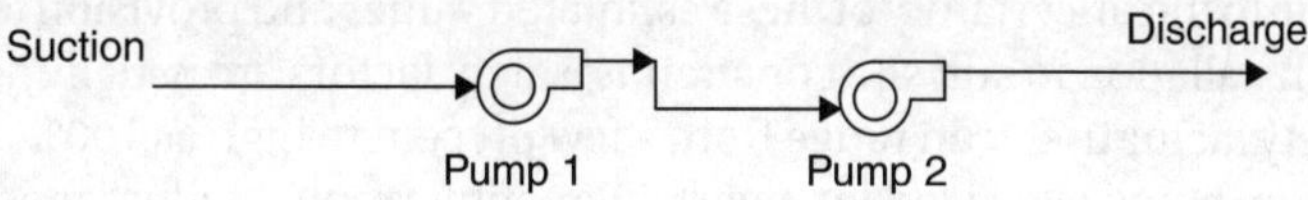

FIGURE 3.10   Arrangement of pumps operating in series.

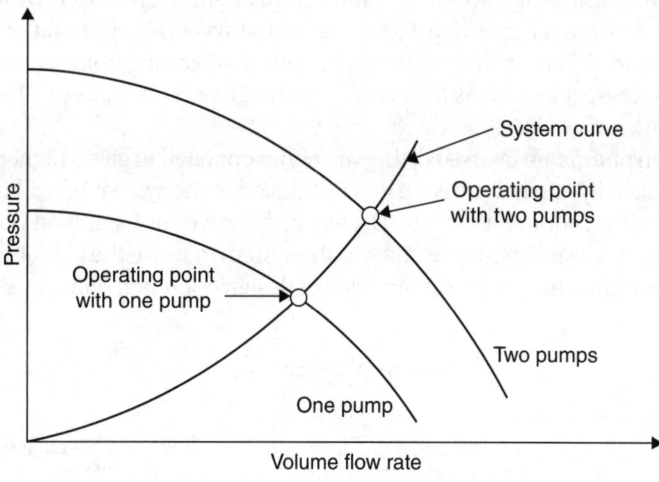

FIGURE 3.11   System curve for series pumps.

3.7   Pump Sizing

When designing pumping systems, it is essential to ensure that the capacity of the pumps selected is sized correctly to prevent energy wastage.

Pumps are sized to provide the design flow requirements while overcoming the various resistances in the system. Friction losses in piping and losses across valves and

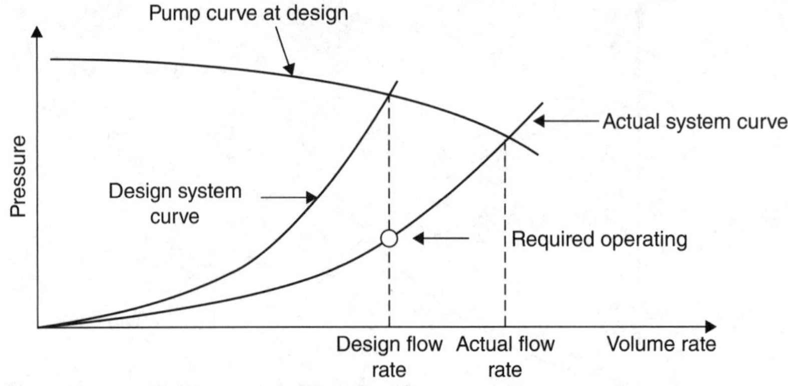

FIGURE 3.12   Oversized pump for application.

fittings are normally estimated based on manufacturer specifications and reference data. Due to the uncertainty of these estimated values and provision for possible changes during installation to suit site constraints, safety factors are usually added to the design. The safety factor used can range from a few percent to high as 100%. As a result, a pump can end up being oversized for a particular application, as illustrated in Fig. 3.12.

The figure illustrates a case where the system curve used at the design stage has a high safety factor incorporated into it. The pump is selected to intersect this design system curve at the design flow to give the design operating point. However, since the design system curve has a high safety factor, the actual system curve that the pump experiences may be quite different. This results in the pump operating point moving along the pump curve to where it intersects the actual system curve. This leads to a higher than required pump flow.

When pumps are oversized, they are either operated to give a higher than required flow or an artificially created pressure loss is induced in the system by adding a throttling valve to achieve the required flow rate. Usually globe valves or balancing valves are used in the system to add a sufficient resistance to the system to move the actual system curve so that it intersects the pump curve at the original design operating point, as shown in Fig. 3.13.

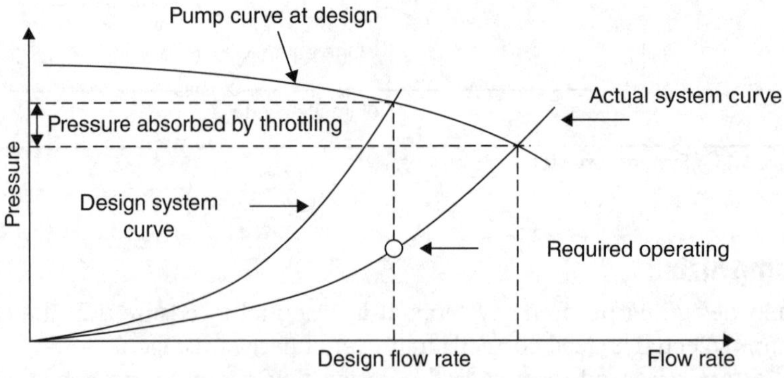

FIGURE 3.13   Oversized pump with throttling.

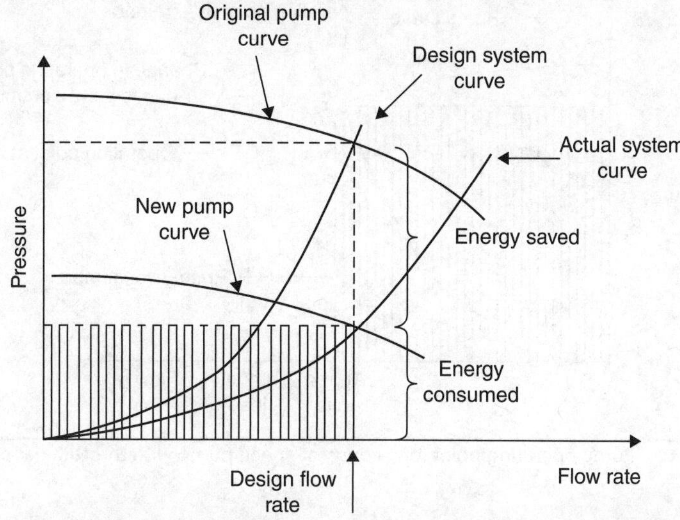

FIGURE 3.14   Energy consumed by the same pump if the impeller diameter or speed is reduced to give the design flow.

Although, operationally, it may be possible to tolerate these options, they should not be accepted from an energy efficiency point of view as a higher than required flow or pressure results in higher pumping power consumption.

In such a situation, the design flow rate can be achieved by reducing the impeller diameter (through trimming the impeller) or reducing the speed of the pump [using a variable-frequency drive (VFD)]. VFDs are also sometimes called variable-speed drives (VSDs) or adjustable-frequency drives (AFDs). These are devices that can convert the frequency of the utility power supply and provide an adjustable output voltage and frequency to vary the speed of motors.

The resulting energy savings due to speed reduction and impeller diameter reduction is illustrated in Fig. 3.14.

Whether to reduce the speed or impeller diameter normally depends on the relative cost for the two options as well as other factors such as the possibility of variable flow applications or possible rise in demand for flow rate in the future. Normally, reducing the pump speed using a VFD is preferable since it can be used to vary the pump speed and capacity when the load changes. Also reducing the impeller diameter can result in a bigger drop in pump efficiency as compared to use of a VFD for reduction of speed. However, if the existing pump is old and due for replacement, the option of replacing the pump with a correctly sized new pump should also be considered.

3.8   Constant-Flow versus Variable-Flow Systems

In variable load systems, pumps are sized to meet the peak load conditions. At part load, often throttling valves are used to reduce the capacity of pumps to match the load conditions. Alternatively, VFDs can be used to reduce the capacity of pumping systems at part load to save energy, as shown in Figs. 3.15 and 3.16.

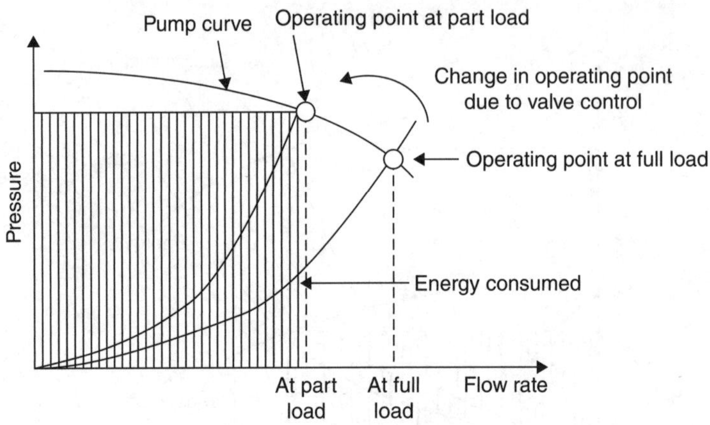

FIGURE 3.15   Pump operating point for a constant-speed pump with throttling at part load.

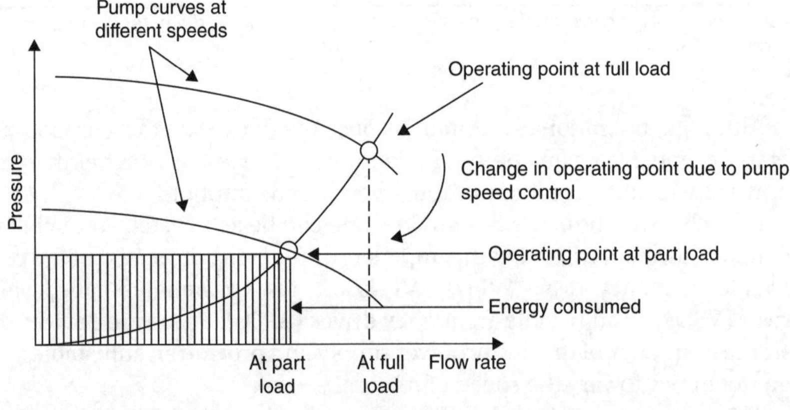

FIGURE 3.16   Pump operating point for a system with variable-speed pumping at part load.

3.9   Effect of Pump Speed and Size on Efficiency

The capacity of pumps depends on the impeller size and operating speed. For a particular operating condition, the same pump may be able to provide the required capacity by using different combinations of impeller sizes and operating speeds. However, as the pump operating efficiency can vary between different selections, the required pump power may also be different for each selection.

Figures 3.17 and 3.18 show two different combinations of impeller sizes and pump speeds, both of which are able to provide a desired pump capacity of 125 L/s and a head of 40 m. As can be seen from the two pump curves, the pump operating at the lower speed requires 66 kW, whereas the other pump requires only 63 kW.

Like pump speed and impeller size, capacity of pumps also affects the operating efficiency. Generally, higher capacity pumps tend to have a better operating efficiency when compared to smaller capacity pumps. Therefore, for a particular application to ensure extra redundancy of pumps, if two pumps are selected each based on 50% of

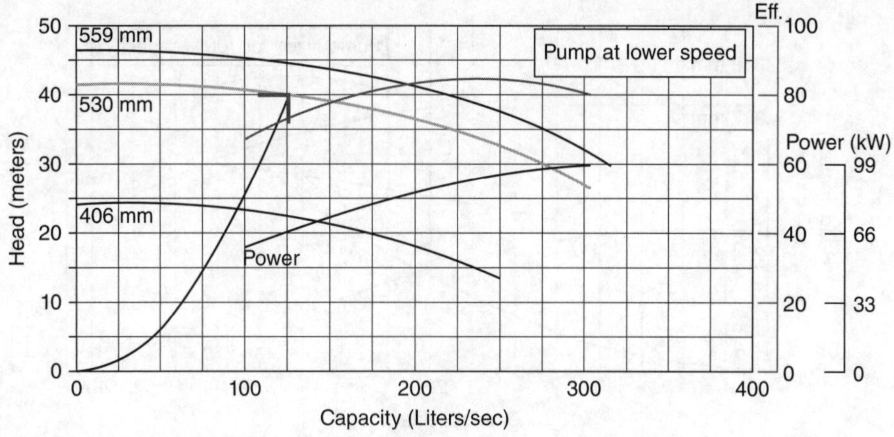

FIGURE 3.17   Pump operating point for a pump operating at a lower speed.

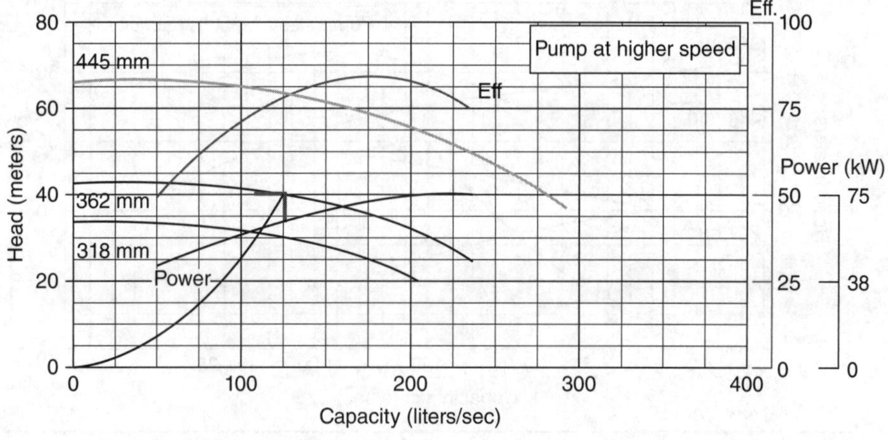

FIGURE 3.18   Pump operating point for a pump operating at a higher speed.

the maximum expected load, instead of one larger pump to meet 100% of the load, the total pump power when operating at 100% load could be higher for the case with two pumps (as illustrated in Figs. 3.19 and 3.20).

Figure 3.19 shows that the operating efficiency for a pump sized for a flow rate of 300 m^3/s and head of 30 m is 84.4%. However, if two pumps are selected for the same application, each with a capacity of 150 m^3/s and head of 30 m (Fig. 3.20), the pump efficiency at the operating point would be about 81%. As a result of the lower efficiency, if one pump is used, the pump power consumption will be 105 kW compared to 110 kW (55 kW × 2) required for two smaller pumps.

Although in the situation described above the higher capacity pump has a better effi-ciency, this may not be the case for all applications. Also, if the application involves a varying pumping load, then having smaller multiple pumps may be better than having one large pump as smaller pumps can be switched off when the load is low.

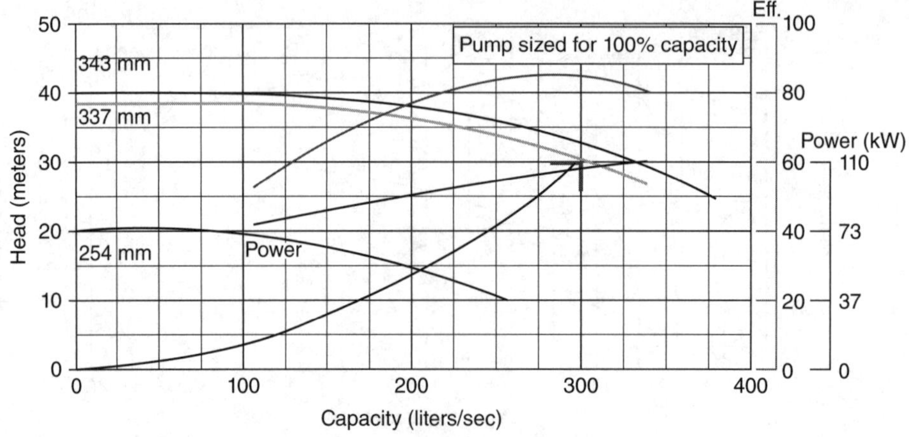

FIGURE 3.19   Pump operating point for a pump sized for 100% of the capacity.

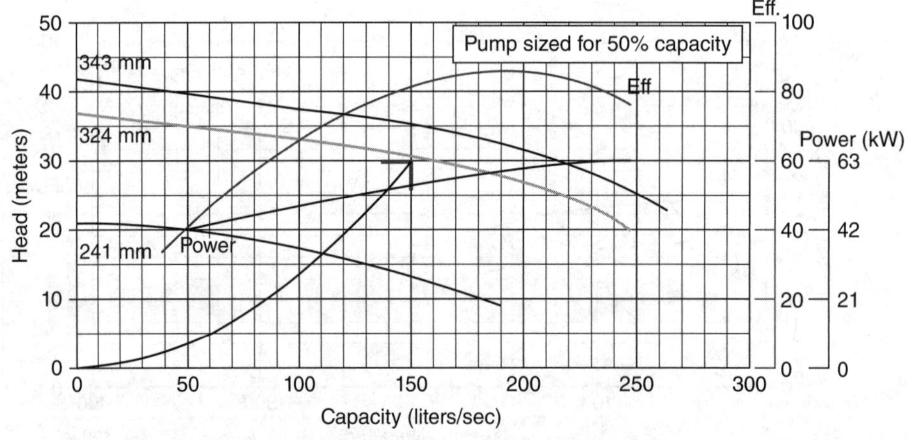

FIGURE 3.20   Pump operating point for a pump sized for 50% of the capacity.

Generally, since the pump operating efficiency is dependent on many factors such as pump type, model, capacity, and operating speed, it is recommended to use pump selection software or consult pump vendors when selecting pumps or designing pumping systems.

3.10   Avoiding Use of Bypass Systems

Often when pumps are oversized for the application, part of the fluid discharged from the pump is circulated back to the suction of the pump through a "bypass" arrangement, as shown in Fig. 3.21. Although such a system is able to provide the design flow requirement to the load, much pumping energy is wasted due to the bypassing action. This energy loss can be eliminated by reducing the capacity of the pump by trimming the impeller diameter or reducing the pump speed so that the pump operating point can be moved to provide the required flow rate.

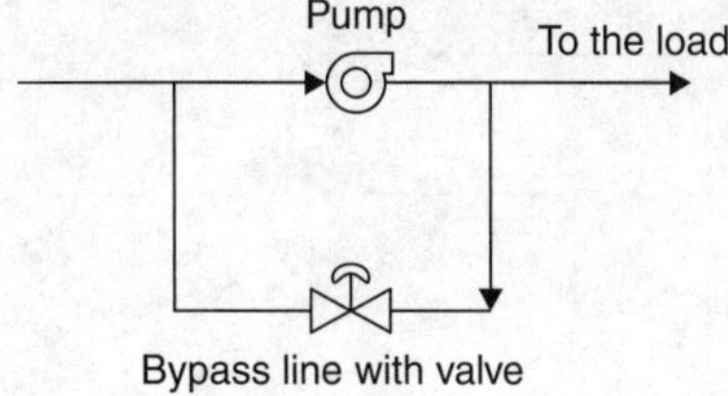

FIGURE 3.21   Pump with a typical bypass arrangement.

3.11   Use of Small Pumps to Augment Larger Pumps

In some systems, the pumping capacity required at peak load and at other times can be significantly different. For example, in sewage systems, the pumping capacity required under normal conditions is much lower than that required during a heavy storm. In such applications, if the pumps installed are selected based on peak demand, the pumps will operate at low efficiency for most of the time.

A possible solution for such situations is to install a small pump, sometimes called a "pony" or "jockey" pump, to handle the normal load while a larger pump is installed to handle the peak load.

3.12   Designing to Minimize Pressure Losses

When a liquid flows through a piping system, head or pressure losses take place due to fluid friction in the piping and resistance offered to the flow by various devices such as valves, strainers, and bends used in the piping system.

The friction losses depend on the pipe material, length of the piping, fluid velocity, and properties of the fluid. Therefore, for a given fluid such as water, the friction losses can be reduced by minimizing the pipe length and reducing the flow velocity (by increasing the pipe diameter).

On the other hand, for a particular flow velocity, losses due to various fittings and devices installed on piping systems depend on the design of the device or fitting. Therefore, for a given flow velocity, different types of valves can have different losses associated with them.

The losses for different types of valves can be significantly different and, therefore, one has to be careful when selecting valves for different applications. For instance, globe-type valves are sometimes used as isolation valves in piping systems. These valves have a high pressure drop even when they are fully open, due to the change in direction the flow has to make when passing through them. On the other hand, butterfly valves when fully open offer little or no resistance to the flow. Therefore, to minimize pumping energy consumption, valves with low resistance (when fully open) such as butterfly valves should be used for flow isolation.

Furthermore, pipe fittings such as bends, elbows, Tees, and flow transition devices should also be selected to minimize head losses in the system, as illustrated in Figs. 3.22 and 3.23.

In piping systems that use "common headers" to serve multiple equipment with different pressure losses, "balancing valves" are installed to ensure that the equipment receives the design flow rate (Fig. 3.24).

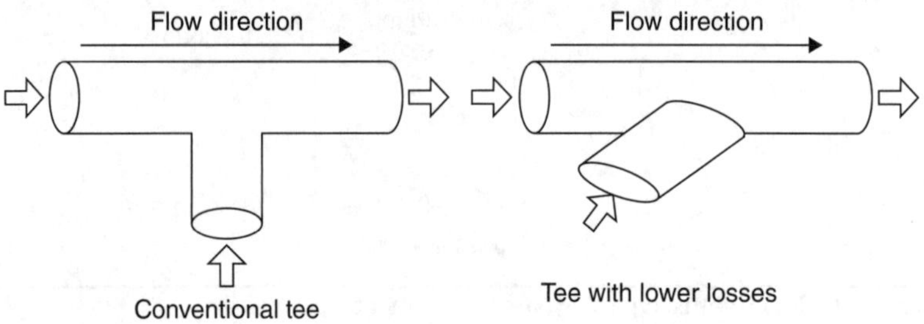

FIGURE 3.22   Design of Tees to reduce pressure losses.

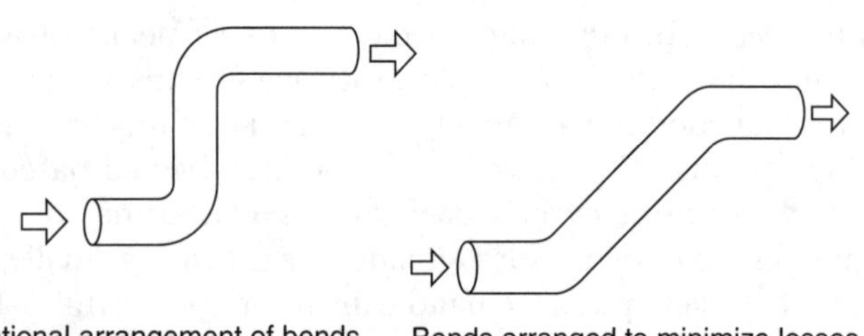

FIGURE 3.23   Design of bends to reduce pressure losses.

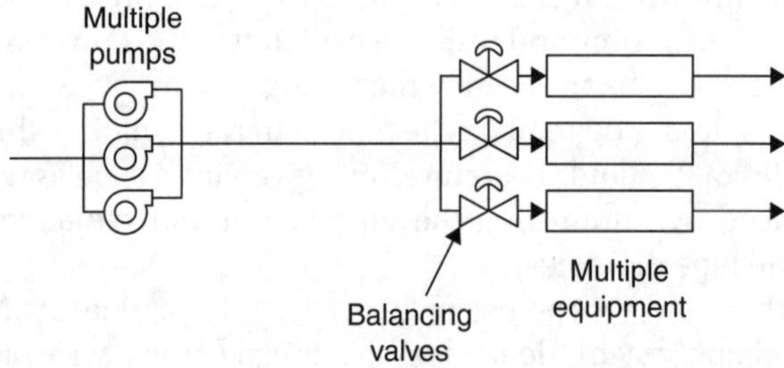

FIGURE 3.24   Arrangement with balancing valves.

Since balancing valves regulate the flow rate by inducing pressure losses to equalize the pressures through the different branches, pumping energy can be reduced by designing the piping arrangement to be a one-to-one system so that the balancing valves can be eliminated, as shown in Fig. 3.25.

3.13   Pump Efficiency

From Eq. (3.2), the pump shaft power can be minimized by selecting a pump with a higher efficiency:

$$\text{Pump impeller power (kW)} = \frac{\text{flow rate (m}^3/\text{s}) \times \text{pressure difference (N/m}^2)}{1000 \times \text{efficiency}}$$

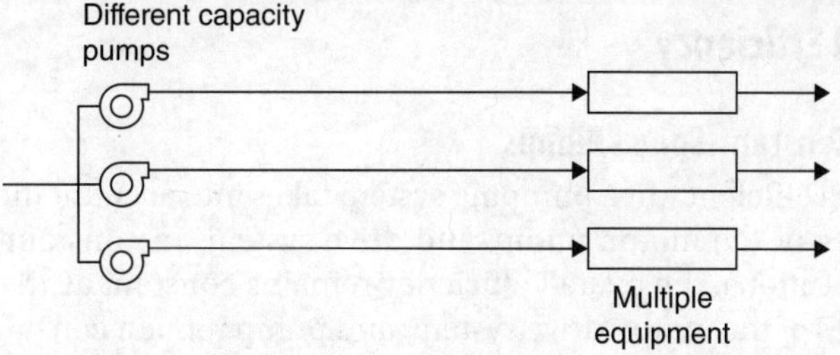

FIGURE 3.25   One-to-one arrangement to avoid flow balancing.

Figures 3.26 and 3.27 show two sets of pump curves for pump A and pump B. Both pumps are able to operate at the desired operating point of flow rate Q and pressure P. However, based on the performance curves, pump A will operate at 90% while pump B will be able to operate at only 78% (estimated by interpolation) at the desired operating point.

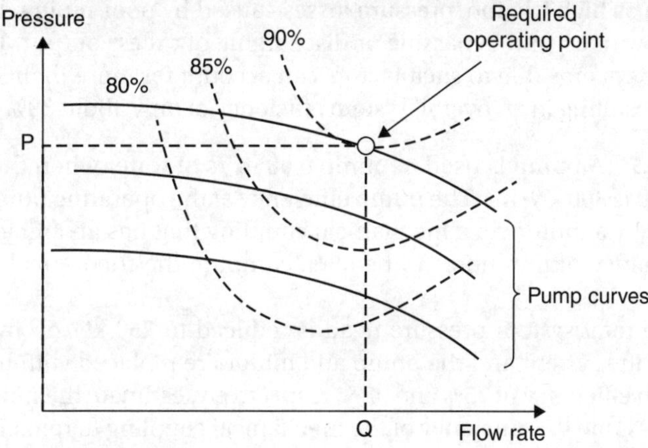

FIGURE 3.26   Operating point for pump A.

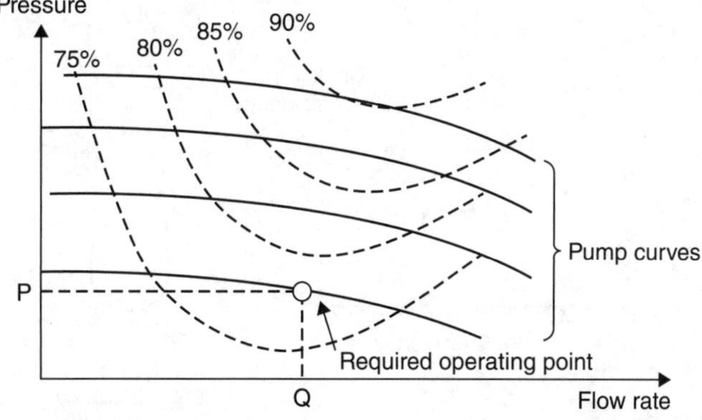

FIGURE 3.27   Operating point for pump B.

3.14   Overall Efficiency

3.14.1   Constant-Speed Pumps

The overall efficiency of a pumping system takes into account the individual operating efficiencies of the motor, pump, and drive system. In constant-speed and constant-load applications, the overall efficiency remains constant as the individual operating efficiencies of the motor, drive system, and pump remain constant.

As shown in Fig. 3.28, for a typical constant-speed application using a standard-efficiency motor of 88%, flexible coupling with a transmission efficiency of 98%, and a pump having an efficiency of 65% at the operating point, the overall energy efficiency of the motor-driven system is 56% ($0.88 \times 0.98 \times 0.65 = 0.56$). This indicates that for every unit of input energy that is used, only 0.56 unit of energy is used to drive the load (0.44 unit of energy is wasted).

In addition to these drive system losses, the pumping system will have various energy losses. As explained earlier, energy wastage in the system can occur due to factors such as high friction/pressure losses caused by poor piping designs, throttling of liquid flow rates, and bypassing or discharging of excess output. The losses in typical pumping systems due to such factors can account for more than 30% of the pumping energy, resulting in an overall system efficiency of only about 39% (Fig. 3.29).

Example 3.5   A pump is used to pump 0.08 m³/s of water when the total system pressure head is 300 kN/m². The pump efficiency at the operating point is 67%. The pump is driven by a motor via a mechanical coupling that has an efficiency of 98%. Taking the efficiency of the motor to be 88%, compute the theoretical input power to the motor.

If the total system pressure head is reduced to 250 kN/m² by reducing pressure losses in the system and the pump and motor are replaced with high-efficiency units having an efficiency of 75% and 92%, respectively, estimate the new input power to the motor. Assume the efficiency of the mechanical coupling to remain the same.

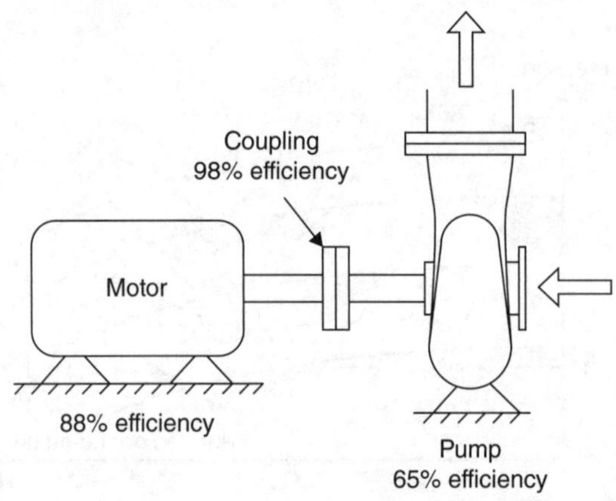

FIGURE 3.28   Overall efficiency of a typical motor-driven pump.

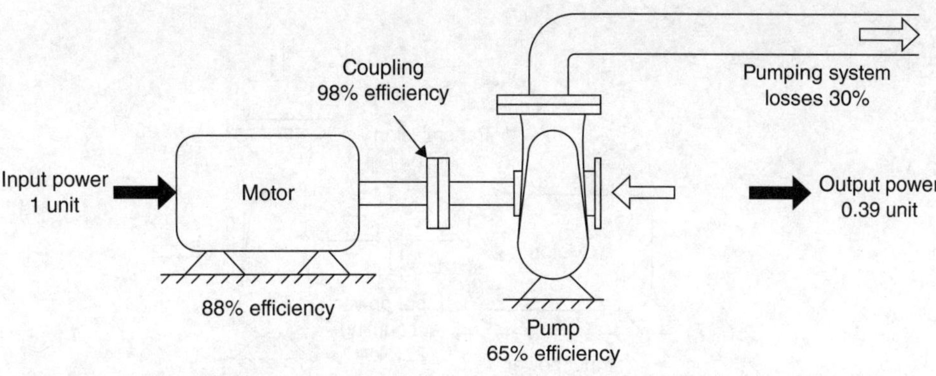

Figure 3.29 Overall energy efficiency of a conventional pumping system.

Solution

Present system:

From Eq. (3.2), the input power to the pump is given by

$$= (0.08 \text{ m}^3/\text{s} \times 300,000 \text{ N/m}^2)/(1000 \times 0.67)$$
$$= 35.8 \text{ kW}$$

Since the mechanical coupling has an efficiency of 98%, the input power at the coupling is 36.5 kW (35.8/0.98). Similarly, the input power to the motor will be 36.5 kW/0.88 = 41.5 kW (see Fig. 3.30).

Proposed system:

From Eq. (3.2), the input power to the pump is given by

$$= (0.08 \text{ m}^3/\text{s} \times 250,000 \text{ N/m}^2)/(1000 \times 0.75)$$
$$= 26.7 \text{ kW}$$

Since the mechanical coupling has an efficiency of 98%, the input power at the coupling is 27.2 kW (26.7/0.98). ▲

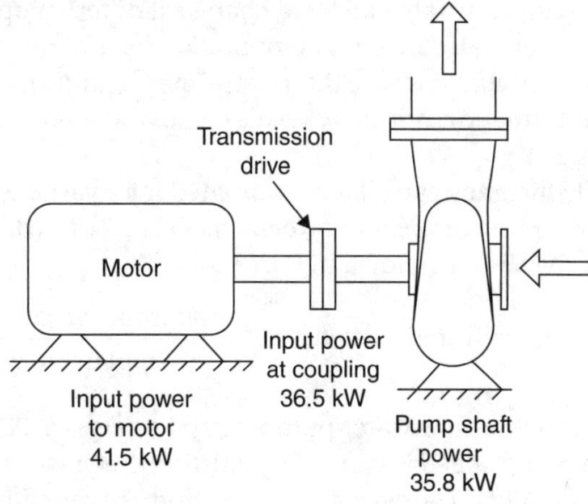

Figure 3.30 Pumping system in Example 3.5.

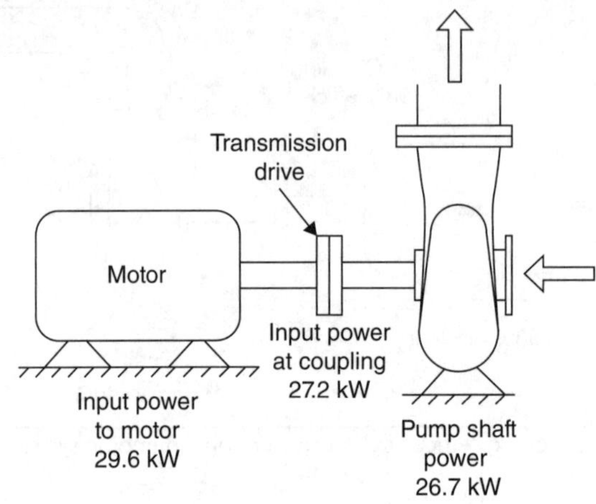

FIGURE 3.31   Proposed pumping system in Example 3.5.

Similarly, the input power to the motor will be 27.2 kW/0.92 = 29.6 kW (see Fig. 3.31). Therefore, the input power has reduced from 41.5 to 29.6 kW, which is a 29% reduction in energy usage.

3.14.2 Variable-Speed Pumps

In variable-load pumping applications where VFDs are used to match the pumping capacity to the load, the overall efficiency does not remain constant as in the case of constant-speed pumps due to the variation of the pump and motor efficiency at different speeds and loads. In addition, in variable-speed applications, the efficiency of the VFD system too needs to be included in the overall efficiency computation.

As illustrated in Chap. 2, the operating efficiency of motors depends on the loading, and the best efficiency is achieved at the rated capacity. Therefore, when a motor experiences lower loads at lower operating speeds, the motor efficiency reduces from its rated value. Similarly, the efficiency of pumps also depends on the operating speed and varies when the flow rate and head change at different operating speeds.

Therefore, the overall efficiency computation for a variable-speed pumping system requires the actual efficiency of the pump, motor, and drive at different operating speeds. Such information normally needs to be obtained from the respective equipment suppliers.

The overall efficiency can also be estimated for a variable-speed pumping system by computing the pump impeller power using Eq. (3.2) and dividing it by the measured power input to the VFD, as shown in Eq. (3.6):

$$\text{Overall efficiency} = \frac{\text{pump impeller power}}{\text{power input to the VFD}} \tag{3.6}$$

Example 3.6   A variable-flow water pumping system uses a VFD to vary the capacity of the pump. The pump speed is varied by adjusting the power supply frequency from 50 to 30 Hz using the VFD. The system flow rate and pressure head at different operating speeds of the pump are provided in Table 3.4. The estimated efficiency of the pump at different operating speeds (obtained from the pump supplier) is given in Table 3.5,

Frequency of power supplied to motor (Hz)	Pump speed (rpm)	System flow rate (m³/s)	System pressure head (kN/m²)
50	1450	0.1	600
45	1305	0.09	486
40	1160	0.08	384
35	1015	0.07	294
30	870	0.06	216

TABLE 3.4   System Flow Rate and Pressure Head

Pump speed (rpm)	Pump efficiency (%)
1450	75
1305	72
1160	70
1015	65
870	62

TABLE 3.5   Pump Efficiency at Different Operating Speeds

whereas the combined operating efficiency for the motor and VFD at different load conditions is provided in Table 3.6. Compute the overall efficiency for the pumping system at the different operating speeds (the efficiency of the drive coupling can be neglected). The motor used is rated at 90 kW.

Solution   From Eq. (3.2), the input power to the pump can be computed in kW as follows:

$$\text{Input power to the pump} = (\text{system flow rate, m}^3/\text{s} \times \text{system pressure, N/m}^2)/(1000 \times \text{pump efficiency})$$

Similarly, pump impeller power (kW) = system flow rate, m³/s × system pressure, kN/m²)   ▲

Motor loading (%)	Motor and VFD efficiency (%)
100	92
90	91
80	88
70	80
60	75
50	70
40	60
30	55
20	50

TABLE 3.6   Combined Efficiency of the Motor and VFD

Pump speed (rpm)	System flow rate (m³/s)	System pressure head (kN/m²)	Pump impeller power (kW)	Pump efficiency (%)	Pump input power (kW)
1450	0.1	600	60.0	75	80
1305	0.09	486	43.7	72	61
1160	0.08	474	30.7	70	44
1015	0.07	294	20.6	65	32
870	0.06	216	13.0	62	21

TABLE 3.7 Estimated Pump Input Power

Pump speed (rpm)	Pump efficiency (%)	Pump input power (kW)	Approximate motor loading (%)	Motor and VFD efficiency (%)	Overall efficiency for the system (%)
	Column A	Column B	Column C = Column B/90 kW	Column D (from Table 3.6)	Column E = Column A × Column D
1450	75	80	89	91	68
1305	72	61	68	80	58
1160	70	44	49	70	49
1015	65	32	36	60	39
870	62	21	23	50	31

TABLE 3.8 Estimated Overall Efficiency for the System

The computed pump input power is tabulated in Table 3.7.

Using the pump input power computed in Table 3.7 and the efficiency of the motor and VFD from Table 3.6 (rounded off to the nearest available value), the respective motor loading and overall efficiencies for the system are tabulated in Table 3.8.

CHAPTER 4

Fan Systems

Different types of fans are used for different applications in industrial plants, including ventilation, removal of contaminants, and industrial processes such as drying.

In such systems, the fan provides the necessary energy to move the air by overcoming frictional losses in the ducting and pressure losses due to components in the system such as filters, dampers, coils, and various fittings. The electrical energy required to operate the system can be minimized if the system design is optimized to reduce these losses.

4.1 Types of Fans

The two main types of fans used for transporting air are centrifugal and axial-flow fans. In centrifugal fans, the rotation of the impeller imparts kinetic energy to the air stream by increasing its velocity. This kinetic energy is later converted to static pressure when the velocity of air is reduced in the diffuser section of the fan prior to discharge. Centrifugal fans are further categorized into different types such as backward curved, forward curved, radial blade, and tubular centrifugal.

In axial-flow fans, the air is pressurized by the aerodynamic lift generated by the fan blades. Axial-flow fans can be further classified as propeller fans, tube-axial fans, and vane-axial fans.

In building and industrial applications, forward-curved and backward-curved centrifugal fans and axial-flow fans are most commonly used (Fig. 4.1).

Forward-curved centrifugal fans are generally less efficient than axial-flow and backward-curved centrifugal fans, but are commonly used for low-static-pressure applications due to their relatively lower cost. Backward-curved centrifugal fans and axial-flow fans are relatively more efficient with highest achievable efficiencies of 90% and 80%, respectively (Table 4.1). Therefore, backward-curved centrifugal fans should be used for low-volume, high-pressure applications, while axial-flow fans should be used for high–volume, low-pressure applications.

4.2 Fan and System Characteristics

Fans and fan systems are normally rated based on pressure and flow rate. The two parameters are dependent on each other as the airflow rate produced by a fan depends on the system pressure (the pressure against which it needs to work).

Figure 4.1 Commonly used fans (courtesy of Flakt Woods).

Fan type	Airflow	Pressure	Efficiency
Axial flow	High	Low	60%–80%
Centrifugal—forward curved	High	Low	50%–70%
Centrifugal—backward curved	Low	High	65%–90%

Table 4.1 Common Types of Fans and Comparison of Performance

The system resistance of a ducting system is the sum of all pressure losses encountered in the system due to fittings, filters, ducting, dampers, and various equipment.

The system curve is a plot of the system resistance encountered at different volumes of airflow. The system resistance varies with the square of the airflow and the curve is parabolic in shape. Similarly, the relationship between the flow and pressure developed by a fan is called a fan curve (Figs. 4.2 and 4.3).

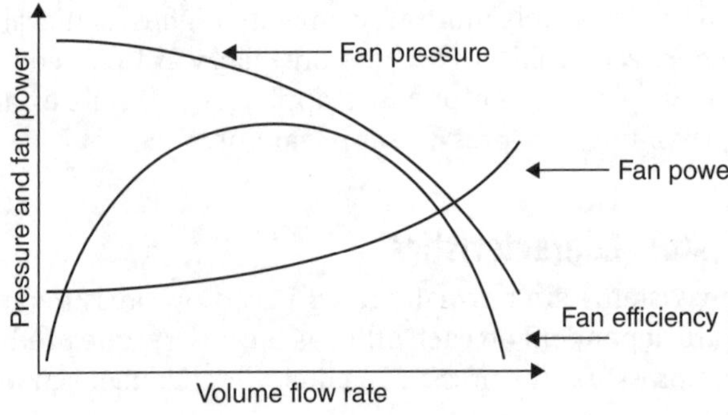

Figure 4.2 Typical fan curves provided by manufacturers.

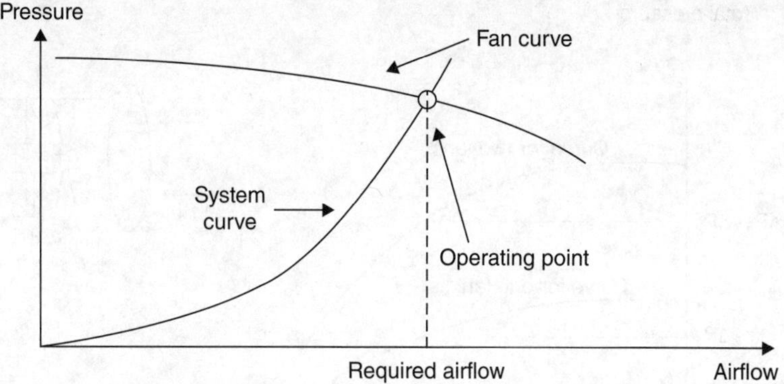

FIGURE 4.3   Matching of system and fan curves.

A fan curve shows all the different operating points of a fan at a particular operating speed as its discharge is throttled from zero to full flow. Since fans can operate at different speeds and with different impeller sizes, usually fan curves for different fan speeds and impeller sizes are plotted on the same axis.

The actual shape of a fan curve depends on the type of fan used. The performance of some fans such as axial-flow fans also depends on the blade angle, which is adjustable. For such fans, manufacturers provide a "family of curves" showing fan performance at different blade settings. Fan curves also include data such as the fan power required and operating efficiency under different operating conditions.

4.3   Fan Selection

The system operating point is the point on the system resistance curve which corresponds to the required airflow condition. A fan selected for a particular application has a performance curve intersecting the system curve at the desired operating point.

Fans can be installed to operate as a single fan or in multiple-fan systems in series or parallel with other fans.

In series fan operation (Fig. 4.4), the total airflow remains the same as the airflow rate for one fan, but the pressure developed is the sum of the pressures developed by the individual fans.

Similarly, in parallel operation (Fig. 4.5), the pressure developed by the fans is the same as the pressure developed by one fan, while the total airflow becomes the sum of the airflow rates for the individual fans.

In applications where the airflow quantity to be transported varies significantly, two or more smaller fans can be installed in parallel in place of one large fan so that the fan capacity (number of fans in operation) can be varied to match the demand. Such a design can help to operate the fans closer to their best efficiency point under different load conditions, rather than modulating the capacity of one large fan, which could lead to lower fan operating efficiency.

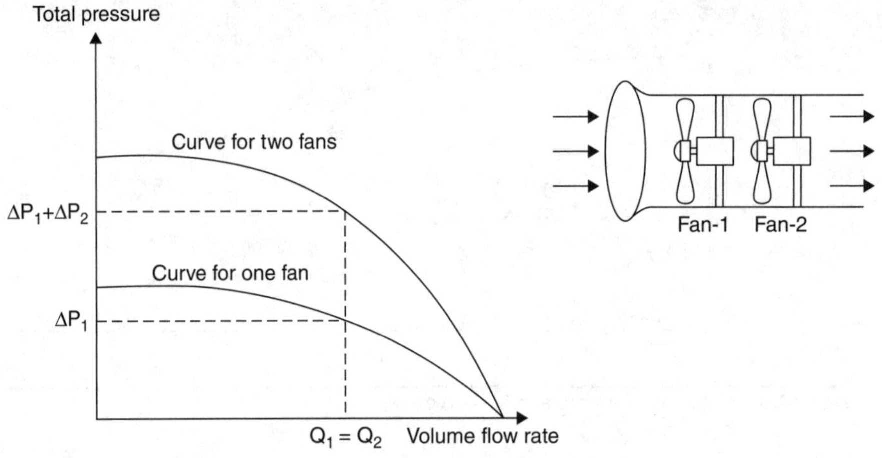

FIGURE 4.4   Operation of fans in series.

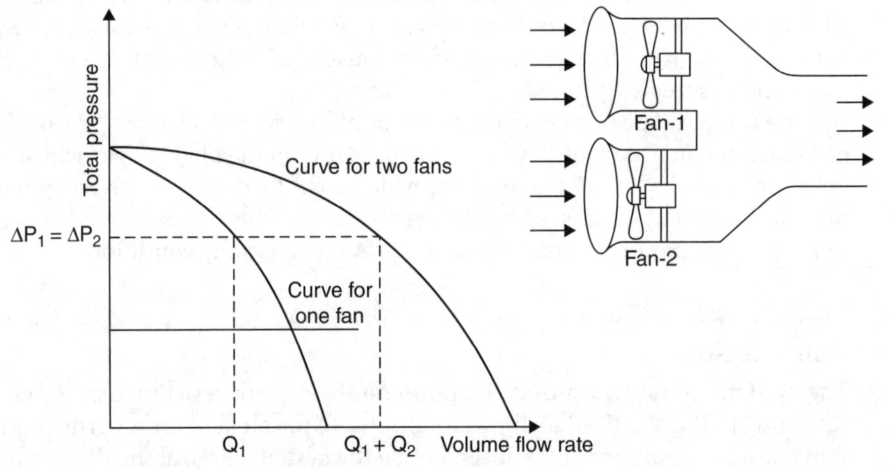

FIGURE 4.5   Operation of fans in parallel.

4.4   Theoretical Fan Power Consumption

The power consumed by fans (fan brake horsepower or impeller power) is proportional to the product of the volume flow and pressure developed:

$$\text{Fan impeller power} \propto \frac{\text{airflow rate} \times \text{pressure developed}}{\text{efficiency}} \tag{4.1}$$

In SI units:

$$\text{Fan impeller power (kW)} = \frac{\text{flow rate (m}^3/\text{s)} \times \text{pressure developed (N/m}^2 \text{ or Pa)}}{1000 \times \text{efficiency}} \tag{4.2}$$

In Imperial units:

$$\text{Fan brake horsepower} = \frac{\text{flow rate (cfm)} \times \text{pressure difference (in. water)}}{6350 \times \text{efficiency}} \quad (4.3)$$

Therefore, based on Eqs. (4.1) to (4.3), fan power consumption can be lowered by reducing the airflow rate, reducing system pressure losses, and improving fan efficiency. These are the main strategies employed to reduce energy consumption of fans and fan systems.

4.5　Affinity Laws

The performance levels of fans under different conditions are related by the fan affinity laws given in Table 4.2. The fan affinity laws relate fan speed and impeller diameter to airflow, pressure developed by the fan, and impeller power.

The fan affinity laws are used by manufacturers to estimate the performance of fans at different impeller sizes and operating speeds. Since the speed of fans can be easily changed by changing the size of the pulley or by installing variable-speed drives, the fan laws are very useful to system designers. The fan laws can be used to estimate the performance of fans at different speeds.

Example 4.1　A fan delivers 4 m³/s at 1450 rpm and consumes 25 kW. Calculate the new airflow rate and power consumption if the fan speed is reduced to 1250 rpm.

Solution

$$Q_1 = 4 \text{ m}^3/\text{s}$$

$$P_1 = 25 \text{ kW}$$

$$N_1 = 1450 \text{ rpm}$$

$$N_2 = 1250 \text{ rpm}$$

$$Q_2 = Q_1 \times (N_2/N_1) = 4 \times (1250/1450) = 3.45 \text{ m}^3/\text{s}$$

$$P_2 = P_1 \times (N_2/N_1)^3 = 25 \times (1250/1450)^3 = 16 \text{ kW} \quad \blacktriangle$$

	Change in speed (N)	Change in impeller diameter (D)
Flow (Q)	$Q_2 = Q_1\left(\dfrac{N_2}{N_1}\right)$	$Q_2 = Q_1\left(\dfrac{D_2}{D_1}\right)^3$
Pressure (Δp)	$\Delta p_2 = \Delta p_1\left(\dfrac{N_2}{N_1}\right)^2$	$\Delta p_2 = \Delta p_1\left(\dfrac{D_2}{D_1}\right)^2$
Power (P)	$P_2 = P_1\left(\dfrac{N_2}{N_1}\right)^3$	$P_2 = P_1\left(\dfrac{D_2}{D_1}\right)^5$

TABLE 4.2　Fan Affinity Laws

This example also shows that when the speed is reduced by 14% (1450 rpm to 1250 rpm), theoretically, the power consumption reduces by about 36% (25 kW to 16 kW). It should be noted that this calculation does not take into account the change in fan efficiency due to change in operating speed.

4.6   System Losses

When air flows in ducting systems, there is a pressure drop due to "friction losses" and "dynamic losses" caused by a change of direction or velocity in ducts and fittings.

Friction losses are due to fluid viscous effects and can be expressed by means of D'Arcy's equation

$$\Delta P_f = \frac{f \times L \times v^2}{2 \times g \times D_m} \tag{4.4}$$

where ΔP_f = frictional pressure drop, f = friction factor, L = length of the duct, v = mean duct velocity, g = acceleration due to gravity, D_m = hydraulic mean diameter

$$= \frac{\text{cross-sectional area}}{\text{perimeter}}.$$

To simplify the task of ducting system design, friction losses per unit length for various duct sizes and materials are generally expressed in the form of charts and tables and are available in reference books such as the *ASHRAE Handbook of Fundamentals*.

In good ducting system designs, to minimize friction losses, the cross-section of ducts is usually selected so as to maintain friction losses in the range of 1 to 2 Pa/m. Table 4.3 provides a list of sizes selected to maintain the friction losses in the region of 1 to 2 Pa/m for galvanized ducts.

Dynamic losses occur due to a change in flow direction caused by fittings such as elbows, bends, and tees, and change in area or velocity caused by fittings such as diverging sections, contracting sections, openings, and dampers.

Airflow rate (L/s)	Recommended round duct diameter (mm)
50	160
200	250
500	315
1,000	400
2,000	630
5,000	800
10,000	1,000
20,000	1,250

TABLE 4.3   Round Duct Sizes for Various Airflow Rates

Normally, dynamic pressure losses (ΔP_d) are proportional to the velocity pressure and can be expressed as

$$\Delta P_d = C_o . P_v \tag{4.5}$$

where C_o = dynamic loss coefficient dependent on the geometry of the particular fitting and P_v = velocity pressure $\left(\dfrac{1}{2}\rho v^2\right)$ (ρ is the density of air).

The value of C_o is measured experimentally and is tabulated in reference books such as the *ASHRAE Handbook of Fundamentals*. Selected values of the dynamic loss coefficient are listed in Figs. 4.6 and 4.7.

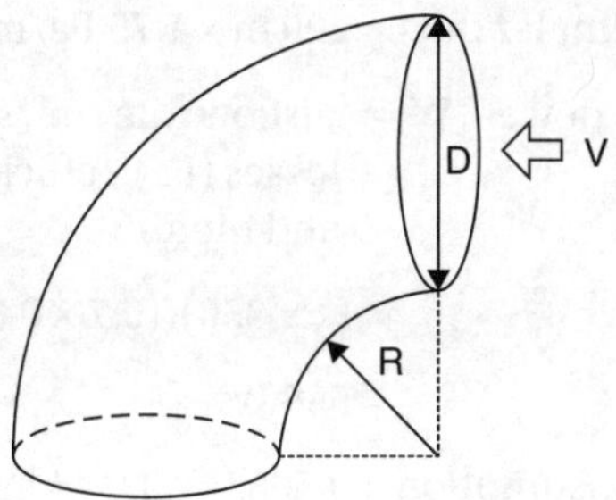

R/D Ratio	C_o
0.25	0.45
0.50	0.34
1.0	0.24
1.5	0.23
2.5	0.22

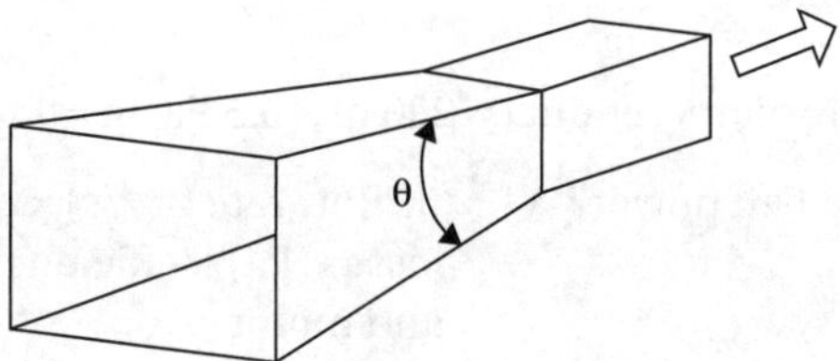

θ	C_o
15	0.008
30	0.02
45	0.04
65	0.07

FIGURE 4.6   Dynamic loss coefficient for 90° bends and duct transformations.

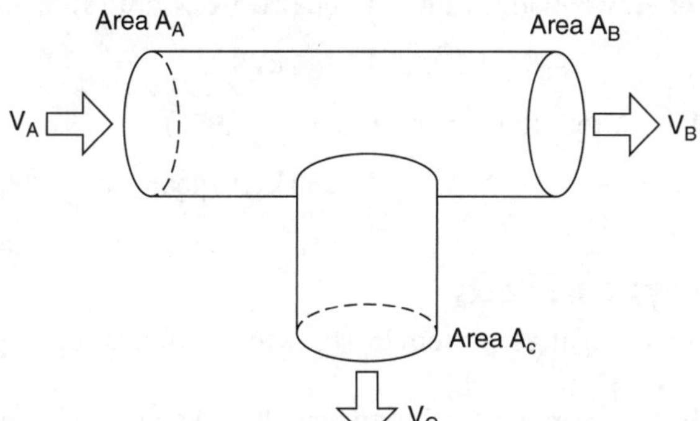

R/D Ratio	C_o
0.25	0.45
0.50	0.34
1.0	0.24
1.5	0.23
2.5	0.22

FIGURE 4.7   Dynamic loss coefficient for straight through Tee fittings.

Example 4.2   A ventilation system which mainly comprises an axial fan and a 200-m-long duct operates 24 hours a day and is designed to transport 1000 L/s of air. If the fan has a rated efficiency of 70% at the operating point, and the duct size is selected based on Table 4.3 (400 mm diameter), compute the annual fan motor energy consumption. Assume motor efficiency to be 88% and the duct pressure loss to be 1.75 Pa/m. Estimate the increase in annual energy consumption that would result if a 315-mm round duct is used instead (based on reference data from the *ASHRAE Handbook of Fundamentals*, the duct pressure losses for the 315-mm round duct with an airflow rate of 1000 L/s is about 5.5 Pa/m).

Solution

$$\text{Total duct losses for a 400-mm-diameter duct} = 200 \text{ m} \times 1.75 \text{ Pa/m} = 350 \text{ Pa}$$

$$\text{Fan power (W)} = [\text{airflow rate } (\text{m}^3/\text{s}) \times \text{pressure losses (Pa)}]/\text{efficiencies of fan and motor}$$

$$= (1 \times 350)/(0.7 \times 0.88)$$

$$= 568 \text{ W}$$

$$\text{Annual energy consumption} = 0.568 \text{ (kW)} \times 24 \text{ hours} \times 365 \text{ days}$$

$$= 4{,}976 \text{ kWh}$$

If a 315-mm-diameter duct is used:

$$\text{Total duct losses for a 315-mm-diameter duct} = 200 \text{ m} \times 5.5 \text{ Pa/m} = 1100 \text{ Pa}$$

$$\text{Fan power (W)} = [\text{airflow rate } (\text{m}^3/\text{s}) \times \text{pressure losses (Pa) }]/\text{efficiencies of fan and motor}$$

$$= (1 \times 1100)/(0.7 \times 0.88)$$

$$= 1786 \text{ W}$$

$$\text{Annual energy consumption} = 1.786 \text{ (kW)} \times 24 \text{ hours} \times 365 \text{ days}$$

$$= 15{,}645 \text{ kWh}$$

$$\text{Increase in fan motor energy use} = (15{,}645 - 4{,}976)$$

$$= 10{,}669 \text{ kWh/year} \quad \blacktriangle$$

4.7   Fan Discharge and Inlet System Effects

Fan inlet and outlet conditions also affect system losses, which can result in higher fan power to satisfy the requirements.

Fans impart dynamic pressure on the air due to centrifugal action. This dynamic pressure has to be converted into static pressure to enable the fan to overcome the system losses. Usually, a minimum duct length is required after the fan to enable this

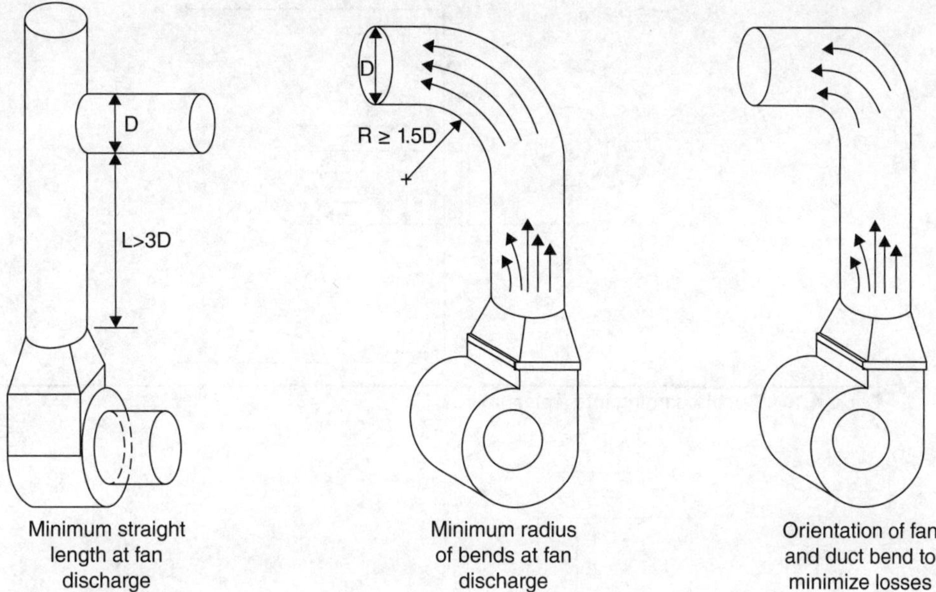

FIGURE 4.8   Minimizing fan discharge losses.

static regain to be completed. Therefore, a system design should attempt to use straight ductwork for 3 to 5 equivalent duct diameters downstream of the fan discharge (Fig. 4.8).

When an elbow must be used at the fan discharge, it is recommended not to have a short-radius elbow. Preferably, an elbow with a minimum radius of 1.5 times the equivalent duct diameter should be used.

Where possible, the orientation of the fan and duct fittings should be arranged to minimize the disturbance to airflow. For example, the bend after the fan can be in the same direction as the airflow profile, as shown in Fig. 4.8.

Where transition to a duct with a larger area is required after the fan, a taper having an included angle of no more than 15° should be used (Fig. 4.9).

Where fans discharge into a plenum (Fig. 4.10), losses occur due to the sudden enlargement in flow area. The addition of a short discharge duct of only 1 or 2 equivalent diameters in length significantly reduces this sudden enlargement loss.

Similarly, fan inlet conditions also affect fan performance. Turbulence at the inlet of a fan reduces the efficiency with which the fan imparts kinetic energy to the

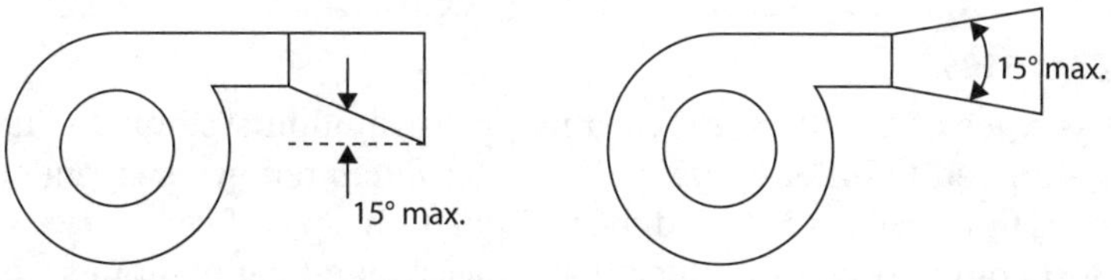

FIGURE 4.9   Fan discharge transitions.

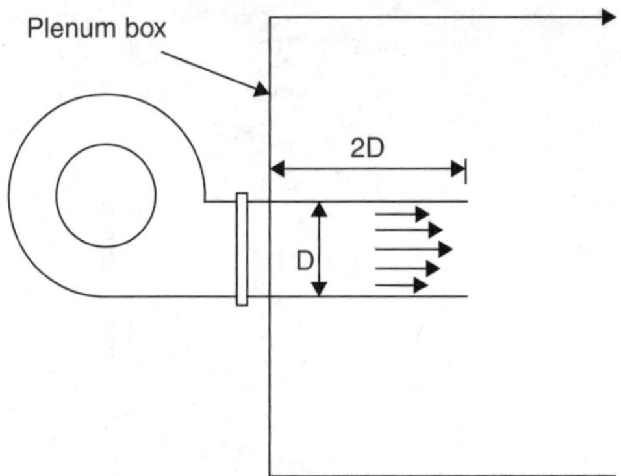

Figure 4.10   Fan discharging into a plenum box.

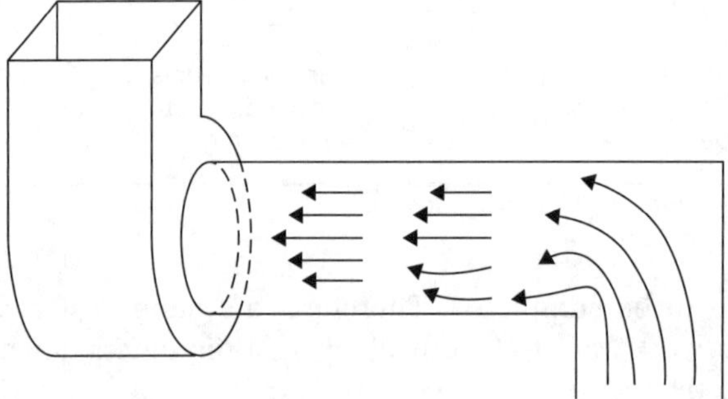

Figure 4.11   Minimum straight duct at fan suction.

air stream. Such nonuniform flow into the suction of a fan is typically caused by an elbow installed too close to the fan inlet. Such an arrangement will not allow the air to enter the impeller uniformly and will cause turbulent and uneven flow distribution at the fan impeller, resulting in lower fan efficiency and higher fan power. Ideally, the minimum straight duct length before the fan should be equivalent to three times the duct diameter (Fig. 4.11). If it is not possible to have a sufficient straight length of ducting at the fan suction, vanes can be installed to minimize losses as shown in Fig. 4.12.

In some applications, the opening is not vertically centered to the fan inlet. By adding a simple splitter sheet on each side of the fan as shown in Fig. 4.13, the performance of the fan can be improved.

4.8   Filter Losses

Various types of air filters are used in air distribution systems to filter the air before being supplied to different users. These air filters remove suspended solid or liquid materials from the fresh air and the recirculation air.

Most common types of filters used for air handling equipment are media filters. These filters are normally made of fibrous material and trap particles when air passes

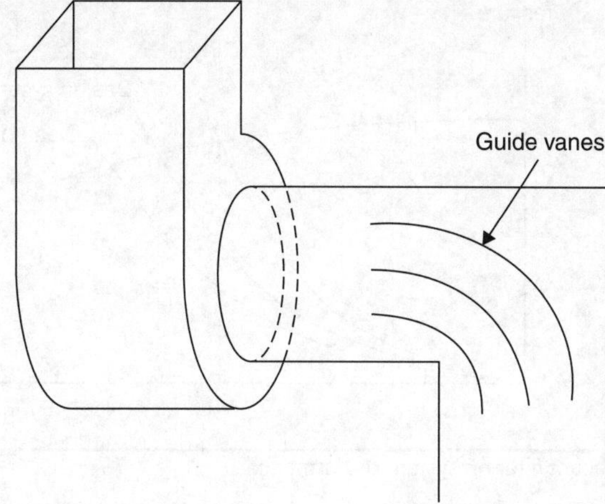

FIGURE 4.12   Use of vanes to straighten airflow.

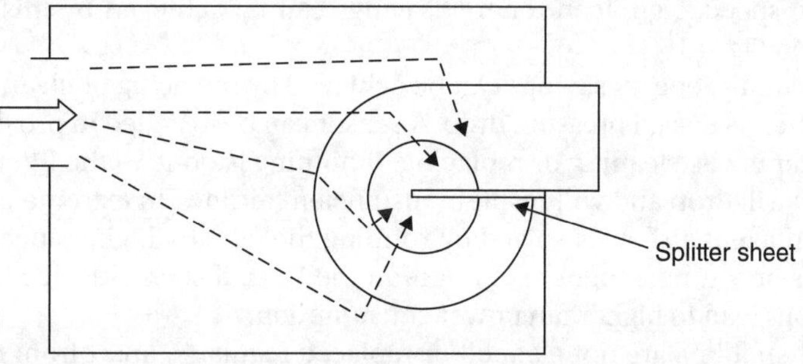

FIGURE 4.13   Splitter sheet for an unbalanced flow.

through them. Media filters offer a resistance to the airflow, depending on the type of filter and the amount of air flowing through it. When these filters accumulate dust, the pressure drop across them increases. Usually, filters are selected to work up to a design pressure drop after which they need to be cleaned or replaced. As shown in the Fig. 4.14, the pressure drop across the filter is lower than the design value when it is clean, which results in a higher airflow (Q_3). Gradually, when the pressure drop increases, the airflow drops to the design value (Q_2). If the filters are not cleaned or replaced at this stage, the higher pressure drop results in lower airflow (Q_1).

From an energy efficiency point of view, there are two aspects that need to be considered in such filter operations. First, when the filter is clean, more airflow than required is provided by the system. From Eq. (4.1) we know that the higher the airflow, the higher the energy consumption, which means that during the period from when the filter is clean to when it reaches the design pressure drop, if the airflow can be reduced, the energy consumed by the fan can be reduced. The easiest way to achieve this is by using a variable-frequency drive (VFD) to modulate the fan speed to provide the design airflow. Since the airflow delivered by the fan is proportional to the

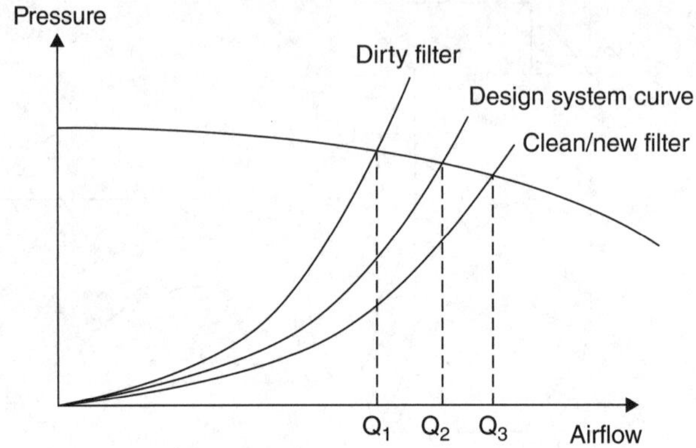

FIGURE 4.14 Effect of filter on system performance.

fan speed, the fan can be operated at speeds lower than the design speed and gradually increased to the design value. Since the fan power is proportional to the cube of the fan speed, significant energy savings can be achieved by this measure as illustrated in Fig. 4.15.

Secondly, energy savings can be achieved by replacing or cleaning the filter when it reaches its design pressure drop. A sensor can be installed to provide an alarm when the filter needs cleaning or replacing. If nothing is done to the filters at this stage, the airflow will drop and will result in insufficient airflow. In extreme cases, this problem of insufficient airflow is solved by running the fan at a higher speed by changing the pulleys or setting a higher set-point on the VFD, if such a device is used. Both these solutions lead to higher fan power consumption.

When filters are not cleaned or replaced regularly, apart from resulting in higher pressure drops, dirt particles sometimes pass through the filters and get lodged in cooling or heating coils. This results in a higher pressure drop across the coils, which leads to higher fan power and the need for more frequent cleaning of the coils, which is even more expensive than cleaning or replacing filters.

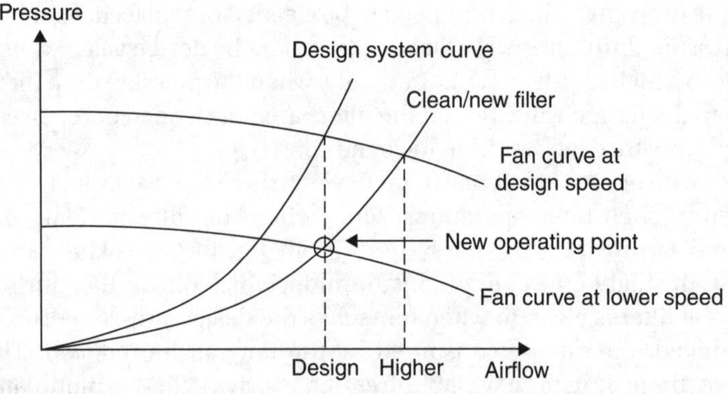

FIGURE 4.15 Effect of reducing fan speed with a clean filter.

Example 4.3   A fan delivers 15 m³/s of air in a system which has a filter with a pressure drop of 200 Pa. Calculate the savings in fan shaft power that can be achieved if this filter is replaced with another which has a pressure drop of only 50 Pa. Assume the total efficiency of the fan to be 65%.

Solution   From Eq. (4.2),

$$\text{Fan impeller power (kW)} = \frac{\text{flow rate (m}^3/\text{s)} \times \text{pressure developed (N/m}^2 \text{ or Pa)}}{1000 \times \text{efficiency}}$$

Therefore, savings in fan power $= [15 \text{ m}^3/\text{s} \times (200 - 50) \text{ Pa}]/(1000 \times 0.65)$

$$= 3.5 \text{ kW} \quad \blacktriangle$$

4.8.1   Face Velocity

In addition to selecting filters with a lower pressure drop, fan power consumption can also be reduced by lowering the velocity of air passing through a filter. Since the pressure drop across a filter is proportional to the square of the airflow velocity (face velocity), a 10% increase in filter face area can reduce the pressure losses by about 20% (reduction in pressure losses due to reduced face velocity is explained further in Sec. 4.9 for cooling and heating coils).

4.9   Coil Losses

4.9.1   Clean Coils

Similar to filters, heating and cooling and heating coils installed in air distribution systems also offer significant resistance. The pressure drop across coils depends on their design (density of rows and fins), face velocity, and how well the coils have been maintained (how clean they are).

Coils are placed downstream of filters so that any particles in the air can be filtered before the air reaches the coil. However, if the filters used are not efficient or not well maintained, they cannot remove dirt particles. This results in the coil acting as the filter. Cooling coils that have been in use for about 10 years and have not been cleaned regularly have a significant drop in airflow due to blockage of the coils. One way of checking the condition of the coil is by comparing the actual coil pressure drop with the coil design pressure drop.

4.9.2   Face Velocity

Apart from keeping the coils clean, fan power consumption can also be minimized through coil design and selection. The pressure drop across coils depends on the number of coil rows and fin density, as well as the face velocity of air flowing through them.

The face velocity of a coil is the velocity of air passing through the coil. The pressure drop across a coil is proportional to the square of the velocity. Therefore, as shown in Fig. 4.16, if the face velocity can be reduced by 10%, the pressure drop can be reduced by about 20%. Reduction in coil face velocity can be achieved by making the coil bigger. However, this leads to a higher first cost for the coil due to its larger size.

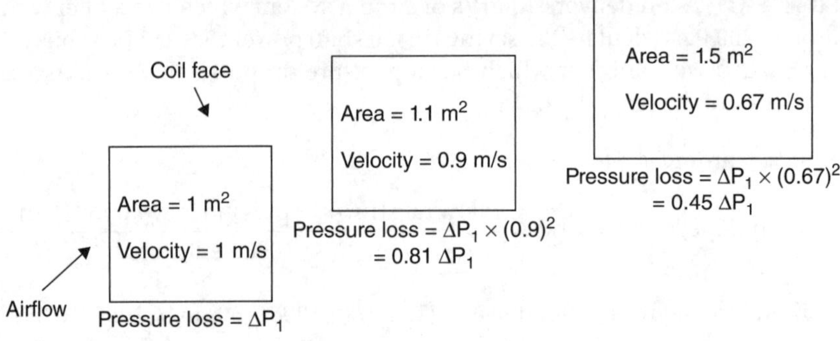

FIGURE 4.16   Coil face area and pressure losses.

Usually, to reduce the first cost, coils are designed for face velocities of about 2.5 m/s. Although coils with low face velocities of about 1 m/s cost more at first, they have a much lower operating cost due to much lower fan energy consumption. Therefore, if coils are selected based on the life cycle cost, low-face-velocity coils offer significant energy savings.

Example 4.4   In a system that delivers 15 m³/s of air, the pressure drop across a cooling coil is 200 Pa and the face velocity is 2.5 m/s. Calculate the pressure drop and the reduction in fan power that will result if the face velocity is reduced to 1.5 m/s. Assume the fan efficiency is 65%.

Solution

$$Q = 15 \text{ m}^3/\text{s}$$

$$v_1 = 2.5 \text{ m/s}$$

$$v_2 = 1.5 \text{ m/s}$$

$$\Delta p_{1(coil)} = 200 \text{ Pa}$$

$$\Delta p_{2(coil)} = \Delta p_{1(coil)} \times (v_2/v_1)^2 = 200 \times (1.5/2.5)^2$$

$$= 72 \text{ Pa}$$

Reduction in pressure drop across the coil $= (200 - 72) = 128$ Pa

From Eq. (4.2), reduction in fan power $= [15 \text{ (m}^3/\text{s)} \times 128 \text{ (Pa)}]/(1000 \times 0.65)$

$$= 2.95 \text{ kW} \quad \blacktriangle$$

4.10   Fan Efficiency

Normally, centrifugal fans used in air distribution systems are either forward curved or backward curved. Forward-curved fans rotate at a relatively slow speed and generally are used for producing high volumes at low static pressure. The maximum static efficiency of forward-curved fans is usually above 70% and occurs at less than 50% of the maximum airflow. The fan power curve has an increasing slope and is referred to as an "overloading type." The advantage of these fans is that they are of relatively low cost.

Backward-curved fans operate at about twice the speed of forward-curved fans. The maximum static efficiency is usually higher than 80% and occurs at about 60% to 70% of the maximum flow (Fig. 4.17). The advantage of backward-curved fans is their higher efficiency and nonoverloading characteristic for the power curve. However, they are normally more costly than forward-curved fans.

Fan efficiency also depends on the operating point, as shown in the Fig. 4.18. Although a backward-curved fan may have a maximum static efficiency of 80%, it may operate at 60% static efficiency at the operating point. Therefore, a fan selected for an application should operate at its highest efficiency at the desired operating point.

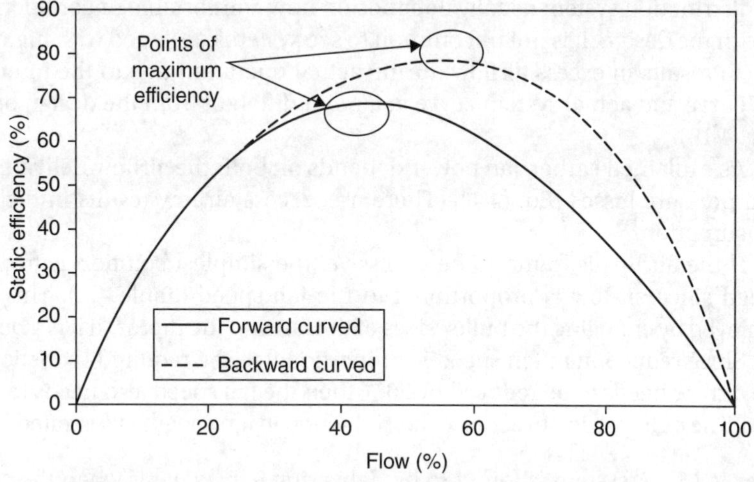

FIGURE 4.17 Fan efficiency for forward-curved and backward-curved fans.

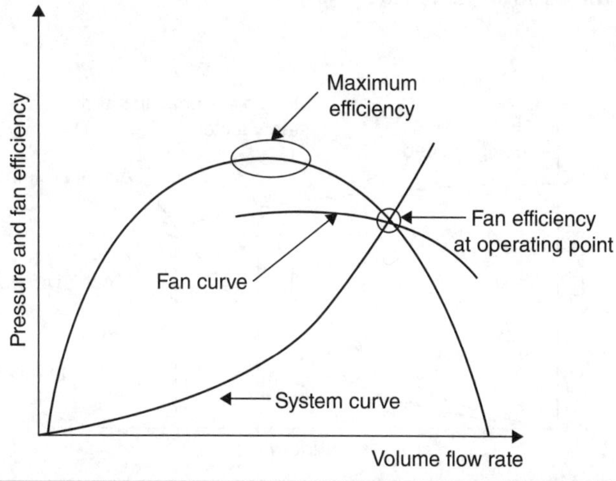

FIGURE 4.18 Fan efficiency at the operating point.

4.11   Right Sizing of Fans

Fan selection for a particular application depends on the expected airflow rate and system pressure losses. The design airflow rate is normally computed based on expected load conditions. Since the airflow delivered by a fan also depends on the pressure losses in the system which it has to overcome to deliver the required airflow rate, the system pressure losses also need to be estimated prior to selecting a fan. As described in Sec. 4.6, system losses depend on factors such as friction losses in ducting, losses due to fittings, and changes in velocity and direction. In system design, usually a safety factor is added to account for differences between computed values and actual system losses that may result due to reasons such as changes in ducting layout during installation to overcome site constraints. The safety factor used for designing an air distribution system usually depends on how comfortable or confident the designer is with the design. It is not uncommon to see systems designed with high safety factors, which results in excess airflow during actual operation due to the intersection of the fan curve and actual system curve at a point different from the design operating point (Fig. 4.19).

As explained earlier, fan power depends on both the airflow delivered and the system pressure losses [Eq. (4.1)]. Therefore, excess airflow results in higher fan power consumption.

If the airflow is found to be excessive, the simplest solution is to reduce the fan speed since airflow is proportional to the fan speed (Table 4.2). This can be easily achieved by changing the pulley sizes of the fan and the motor (if it is a belt-driven fan).

Since reduction in fan speed is proportional to the required reduction in airflow, if the airflow needs to be reduced by 20%, then the fan speed also needs to be reduced by 20%. The pulley sizing to achieve this reduction in fan speed is illustrated in Example 4.5.

Example 4.5   A system is found to be delivering 9 m³/s of air when the actual requirement is 7.5 m³/s. The fan speed is measured to be 750 rpm, while the motor speed is 960 rpm. The motor pulley size is 200 mm. Find the fan speed and pulley size required to reduce the airflow rate to 7.5 m³/s.

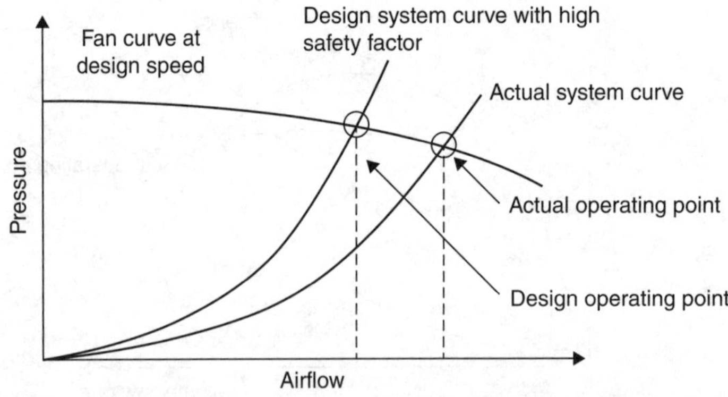

FIGURE 4.19   Fan performance due to the use of a high safety factor for estimating system losses.

Solution

$$Q_1 = 9 \text{ m}^3/\text{s}$$

$$Q_2 = 7.5 \text{ m}^3/\text{s}$$

$$N_1 = 750 \text{ rpm}$$

New fan speed, $N_2 = N_1 \times (Q_2/Q_1) = 750 \times (7.5/9) = 625$ rpm

Since the fan and motor speeds and pulley diameters are related as follows:

$$N_{(fan)}/N_{(motor)} = D_{(motor)}/D_{(fan)}$$

New fan pulley diameter, $D_{(fan)} = [D_{(motor)} \times N_{(motor)}]/N_{(fan)}$

Therefore, $D_{(fan)} = 200 \times (960/625) = 307.2$ mm   ▲

Another common way of reducing the fan speed is by using a VFD for the fan motor and then running the motor at a lower speed. Although this solution is normally more expensive than a pulley change, it is easier to achieve the desired airflow through a trial-and-error setting of VFD frequency.

4.12   Modulating Airflow Rate

From Eq. (4.1), fan power consumption can be reduced by lowering the airflow rate. In many fan systems, the load is variable and peak load conditions are usually experienced only for short periods of time. Therefore, the capacity of fans can be controlled to match load requirements by varying the amount of air supplied. In such variable air volume systems, devices such as discharge dampers, inlet guide vanes, or variable-speed drives are used to regulate the air volume in response to changes in load.

Dampers, which are installed after the fan, increase or decrease the pressure induced by them by opening or closing to create a change in system pressure, which

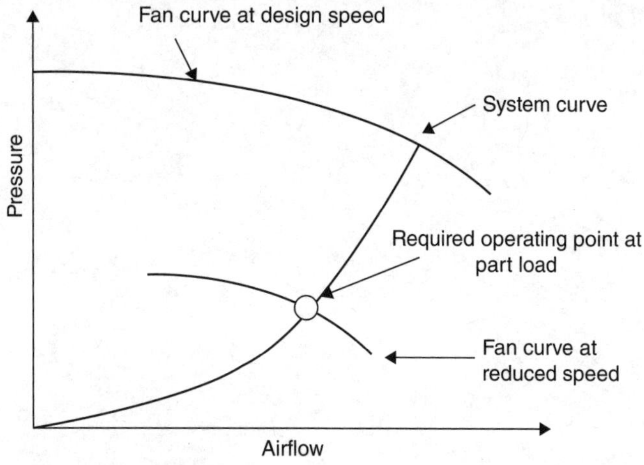

FIGURE 4.20   Reducing airflow at part load by reducing fan speed.

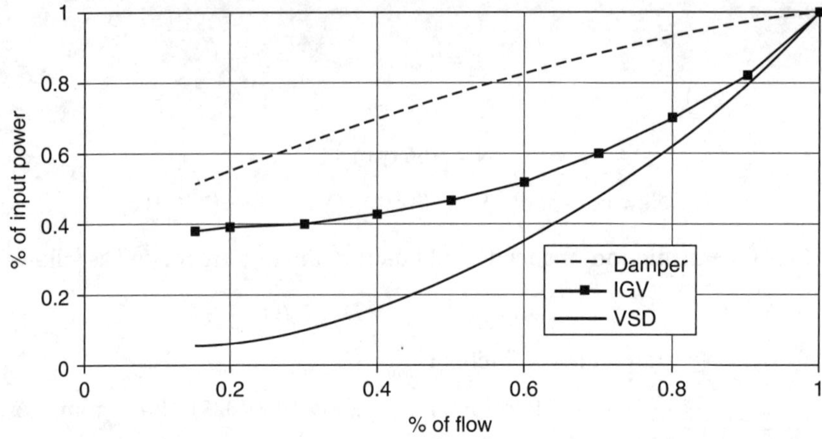

Figure 4.21   Relative fan power consumption for the different flow control systems.

results in the fan operating point moving along the fan curve. Similarly, inlet guide vanes (IGV) installed at the suction of fans, which can be opened or closed to various positions, can also be used to vary the fan output in response to changes in load.

A more energy efficient means of controlling fan capacity is to change the fan speed so that the fan curve can be changed to intersect the system curve at the desired operating point, as shown in Fig. 4.20.

The relative fan power consumption for the three different flow control systems is illustrated in Fig. 4.21.

CHAPTER 5

Boilers and Steam Systems

5.1 Introduction

Steam (water vapor) is used in many industrial applications as the working fluid for heating and as the carrier of energy. Steam is widely used in industry because of easy access to water required for steam generation, the ability to convert water into steam using boilers, the high heat-carrying capacity of steam, and the simplicity in transferring and distributing heat energy through a piping network.

Figure 5.1 shows a typical steam distribution system where steam generated in a boiler is used for jacket heating, process heating, and in sparge coils and ejector systems. The steam supplied for jacket heating and process heating in heat exchangers is condensed and then normally returned to the boiler, unlike in applications such as sparge coils and ejectors where the steam is directly injected into the system.

5.2 Fundamentals of Steam

Figure 5.2 shows the relationship between temperature and specific enthalpy (hereinafter referred to as enthalpy). When water is heated, its temperature rises steadily until it reaches a point called the saturation point where water cannot exist in the liquid form. At this point, addition of heat results in the boiling of water, producing steam. Further heating results in more steam being generated at the same temperature (saturation temperature). When all the water has been converted into steam, further addition of heat leads to an increase in steam temperature above the saturation temperature and the steam is then superheated.

When the evaporation of water is not complete, steam contains moisture or particles of water and is called wet steam. Further addition of heat evaporates all the water in the steam mixture and the steam becomes dry saturated steam. If the dry steam is further heated at a constant pressure, its temperature rises above the saturation temperature and becomes superheated steam.

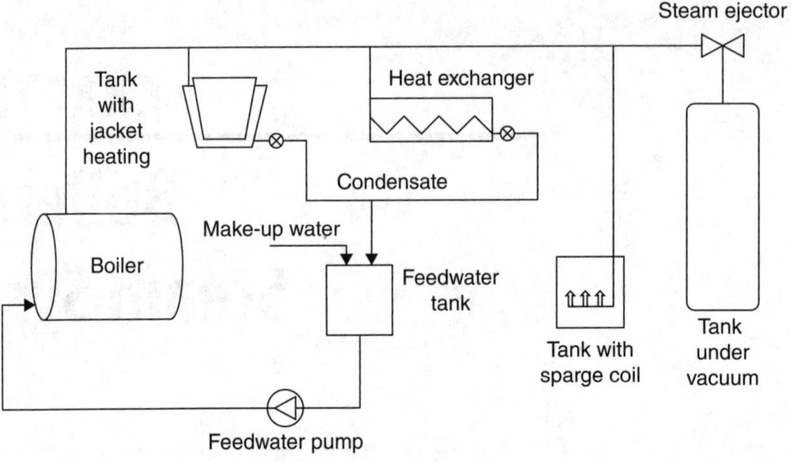

FIGURE 5.1   A typical steam distribution system.

5.2.1 Dryness Fraction of Steam

Dryness fraction of steam is the ratio of the mass of actual dry steam to the total mass of wet steam and can be expressed as

$$x = m_g / m_g + m_f \tag{5.1}$$

$$x = m_g / m$$

where x = dryness fraction, m_g = mass of dry steam, m_f = mass of water in the mixture, and m = mass of wet steam = $m_g + m_f$.

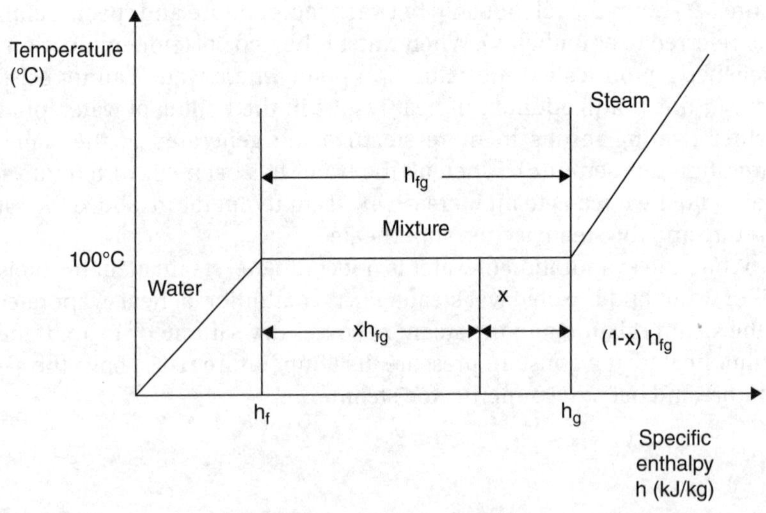

FIGURE 5.2   Conversion of water into steam.

5.2.2 Sensible Heat of Water

Sensible heat is the amount of heat that is required to increase the temperature of 1 kg of water from 0°C to the boiling temperature. It is also called the "liquid enthalpy" or enthalpy of water and is denoted by the symbol h_f. At atmospheric pressure, h_f of water is 419 kJ/kg.

5.2.3 Latent Heat of Water

Latent heat is the amount of heat that is required to change the state of water into steam at the same temperature. The amount of heat required to evaporate 1 kg of water at its boiling point is termed the "enthalpy of evaporation" and is denoted by the symbol h_{fg}. At atmospheric pressure, h_{fg} of water is 2257 kJ/kg.

5.2.4 Total Enthalpy of Steam

Total enthalpy of steam denoted by the symbol h_g is the sum of the sensible heat and latent heat. Therefore,

$$\text{Enthalpy of wet steam } h_g = h_f + x\, h_{fg}$$
$$\text{Enthalpy of dry steam } h_g = h_f + h_{fg}$$

5.2.5 Steam Tables

The properties of steam such as liquid enthalpy, enthalpy of evaporation, and total enthalpy at different pressures and temperatures can be found in "steam tables." An extract of a steam table is provided in Table 5.1.

Absolute pressure (bar)	Temperature (°C)	Enthalpy of water, h_f (kJ/kg)	Enthalpy of evaporation, h_{fg} (kJ/kg)	Enthalpy of steam, h_g (kJ/kg)
1	99.6	417	2258	2675
1.01 (atmospheric pressure)	100	419	2257	2676
2	120.2	505	2202	2707
3	133.5	561	2164	2725
4	143.6	605	2134	2739
5	151.8	640	2109	2749
6	158.8	670	2087	2757
7	165.0	697	2067	2764
8	170.4	721	2048	2769
9	175.4	743	2031	2774
10	179.9	763	2015	2778
11	184.1	781	2000	2781
12	188.0	798	1986	2784

TABLE 5.1 Sample "Steam Table"

5.3   Boilers

Boilers are used to produce steam by adding heat to water. A boiler generally consists of a combustion chamber, which burns fuel in the form of solid, liquid, or gas to produce hot combustion gases, and a tubular heat exchanger to transfer heat from the combustion gases to the water.

As shown in Fig. 5.3, the main inflows to a typical boiler are fuel, air, and feedwater, while the outflows are steam or hot water, exhaust flue gases, and blowdown.

Boilers are normally classified as fire tube or water tube boilers, depending on the flow arrangement of water and the hot gases inside the boiler. In fire tube boilers, the hot gases pass through the boiler tubes that are immersed in the water being heated, while in water tube boilers, water is contained in the tubes which are surrounded by the hot combustion gases.

Fire tube boilers are used for general applications. However, since water is contained in the circular boiler shell in fire tube boilers, an increase in pressure requires an increase in thickness of the boiler shell which becomes impractical at high pressures. Therefore, water tube boilers where water circulated inside tubes (relatively smaller diameter of the tubes compared to the shell diameter in fire tube boilers requires less wall thickness) are used for high-pressure applications.

To increase the surface area available for heat transfer between the combustion gases and water, the tubes in boilers are arranged to have a number of passes so that the hot flue gases can pass through a number of sets of tubes before being exhausted. A typical arrangement of a four-pass fire tube boiler is shown in Fig. 5.4.

Cut-away of a commercial water tube boiler is shown in Fig. 5.5, while a water circulation arrangement is shown in Fig. 5.6. Feedwater enters the steam drum and moves down through the downcomer to the mud drum due to its higher density. This causes the warmer water to rise through the front tubes to the steam drum. The heat added causes some of the water to boil and the steam bubbles are separated and removed at the steam drum.

Boiler capacity depends on the steam generation rate and steam pressure. The steam generation rate is usually rated in kg/h, lb/h, or tons/h, while the steam pressure is rated in psi (pounds per square inch) or bar.

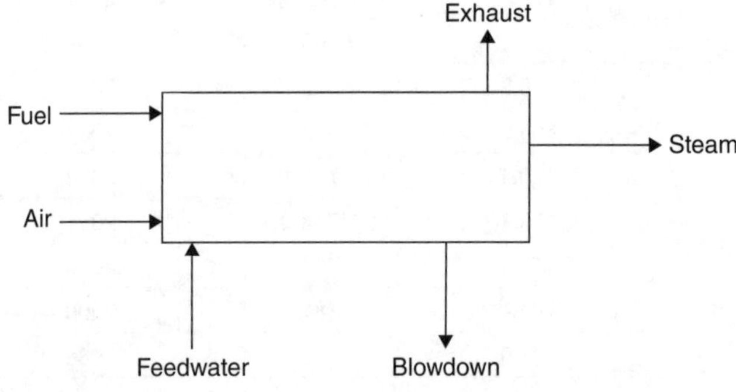

FIGURE 5.3   Main inflows and outflows for a typical boiler.

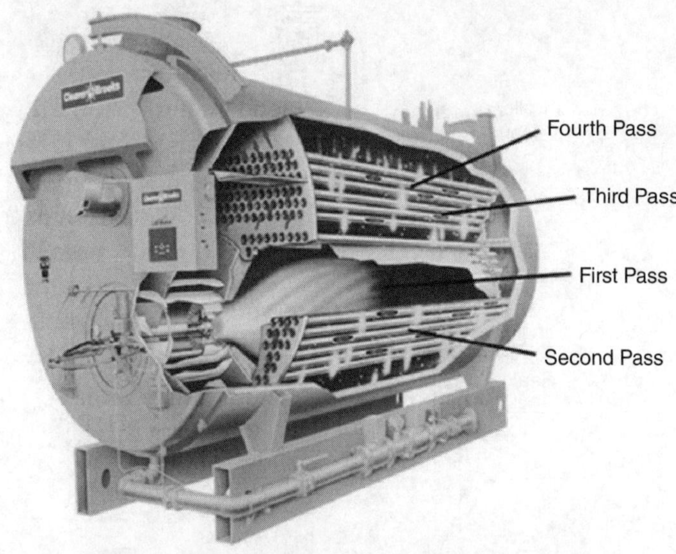

FIGURE 5.4   Arrangement of a four-pass fire tube boiler (courtesy of Cleaver-Brooks).

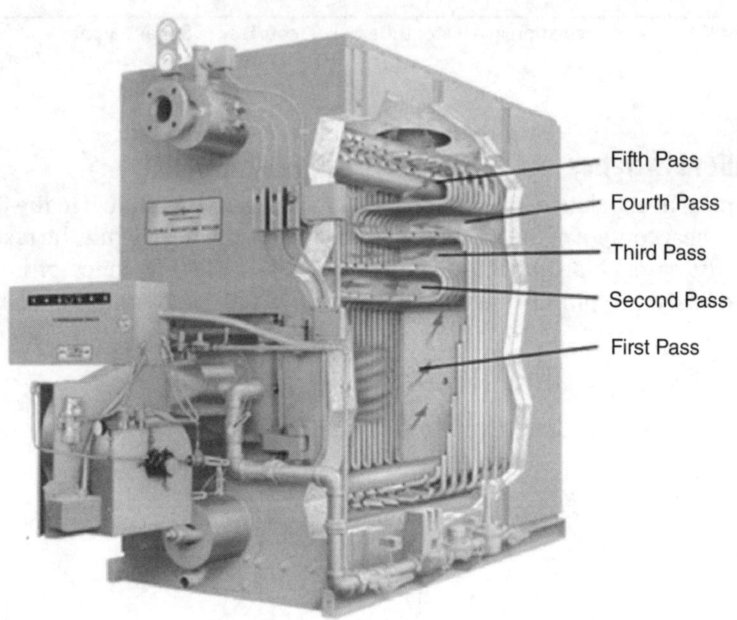

FIGURE 5.5   Cut-away of a commercial water tube boiler (courtesy of Cleaver-Brooks).

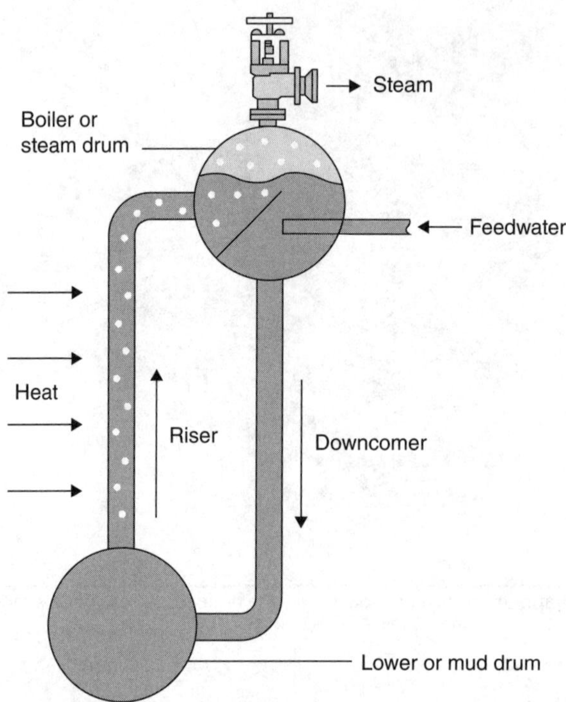

Figure 5.6 Water circulation in a water tube boiler (courtesy of Spirax Sarco).

5.4 Boiler Efficiency

A typical heat balance for a boiler is shown in Fig. 5.7. As shown in the figure, only part of the heat content of the fuel is converted into useful heat, while the rest is lost through exhaust gases, blowdown, and radiation losses. The efficiency of boilers is usually rated based on combustion efficiency, thermal efficiency, and overall efficiency.

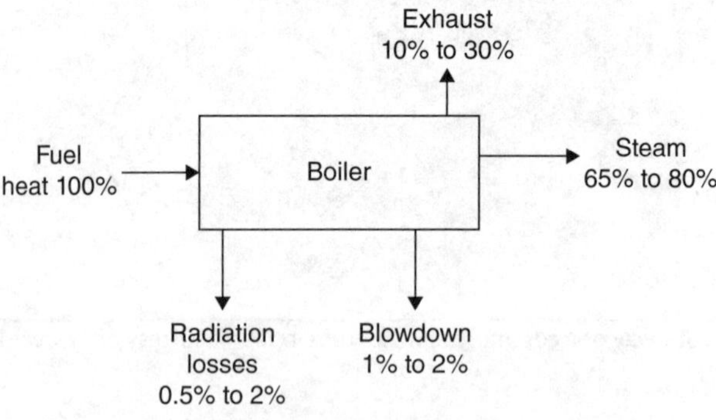

Figure 5.7 Typical heat balance for a boiler.

5.4.1 Combustion Efficiency

A typical combustion process in boilers involves burning of fuels containing carbon (oil, gas, coal) with oxygen to generate heat. Oxygen required for combustion is normally taken from air supplied to the burner of the boiler. The amount of air needed for combustion depends on the type of fuel used. To ensure complete combustion of fuel, more air than required (excess air) for combustion is provided to ensure that the fuel is completely burnt. Since excess air leads to lower boiler efficiency (due to removal of heat by the excess air as it passes through the boiler), the objective is to ensure that an optimum amount of excess air is provided.

One of the most common measures of boiler efficiency is combustion efficiency, which indicates the ability of the combustion process to burn the fuel completely. It is normally measured by sampling the exhaust flue gas to find the composition and temperature using a combustion analyzer.

5.4.2 Thermal Efficiency

Thermal efficiency is a measure of the efficiency of the heat exchange in the boiler. It provides an indication of how well the heat exchanger can transfer heat from the combustion process to water or steam in the boiler. It does not take into consideration the conduction and convection losses from the boiler.

5.4.3 Overall Efficiency

Another measure of boiler efficiency is the overall boiler efficiency, which is a measure of how well the boiler can convert the heat input from the combustion process into steam or hot water. It is also called fuel-to-steam efficiency.

$$\text{Overall boiler efficiency} = \frac{\text{Heat output}}{\text{Heat input}}$$

The heat input depends on the amount of fuel burnt and its calorific value (heating value). The calorific value, normally expressed in kJ/kg, multiplied by the amount of fuel burnt, in kg/s, gives the heat input in kJ/s (kW).

The heat output is the difference between the heat content of feedwater and steam (or hot water) produced multiplied by the flow rate of water or steam. The heat content of water and steam is expressed in kJ/kg, and the flow rate of water or steam is expressed in kg/s, which yields the heat output in kW.

The overall efficiency of a boiler is lower than the combustion efficiency, taking into account radiative and convective losses from the boiler and other losses, such as cycle losses, due to passing of air through the boiler during the "off" cycle.

There are two basic methods of measuring boiler efficiency: "direct method" and "indirect method."

(1) ***Direct method:*** This method is used where the energy gain of the working fluid (water and steam) is compared to the energy content of the fuel used. The energy input is computed based on the calorific value of fuel and amount of fuel used, while the energy output is computed based on the amount of steam generated and the heat content of the feedwater and steam.

For this method, the parameters to be measured are the quantity of steam generated (Q) in kg/h, the quantity of fuel used (q) in kg/h, the pressure and temperature of steam generated (to find h_g of steam), the temperature of feedwater (to find h_g), and the gross calorific value of the fuel used (GCV) in kJ/kg.

$$\text{Boiler efficiency } (\eta) = Q \times (h_g - h_f)/q \times GCV \qquad (5.2)$$

where h_g = enthalpy of saturated steam in kJ/kg of steam and h_f = enthalpy of feedwater in kJ/kg of water.

(2) ***Indirect method:*** In this method, boiler efficiency is computed by subtracting the percentage values of the various losses from 100.

The major heat losses that occur in boilers and are subtracted to estimate the efficiency are as follows:

- Heat loss due to dry flue gas
- Heat loss due to moisture in fuel and combustion air
- Heat loss due to combustion of hydrogen
- Heat loss due to radiation and unaccounted loss
- Heat loss due to unburnt carbon in fly ash and bottom ash

$$\text{Boiler efficiency } (\eta) = 100 - (A + B + C + D + E) \qquad (5.3)$$

Example 5.1   The following data were recorded during a boiler test:

$$\text{Mass of solid fuel used} = 280 \text{ kg}$$
$$\text{Water evaporated} = 2450 \text{ kg}$$
$$\text{Steam pressure} = 12 \text{ bar (absolute)}$$
$$\text{Dryness fraction of steam} = 0.97$$
$$\text{Feedwater temperature} = 52°C$$

The calorific value of the fuel used is 28,500 kJ/kg and enthalpy of feedwater (from steam tables) is 217.6 kJ/kg. Compute the boiler efficiency using the direct method.

Solution   From the steam tables, the enthalpy of steam generated at 12 bar is as follows:

$$h_f = 798 \text{ kJ/kg}$$
$$h_{fg} = 1986 \text{ kJ/kg}$$

Enthalpy equation for wet steam, $h_g = h_f + x h_{fg} = 798 + (0.97 \times 1986) = 2724 \text{ kJ/kg}$

$$\text{Boiler efficiency} = Q \times (h_g - h_f)/q \times GCV \text{ [from Eq. (5.2)]}$$
$$= [2450 \times (2724 - 217.6)]/(280 \times 28{,}500)$$
$$= 0.77$$
$$= 77\% \quad \blacktriangle$$

5.5   Energy-Saving Measures for Boiler Systems

5.5.1   Improving Combustion Efficiency

The major loss in any boiler is due to the hot gases discharged into the chimney. If there is a lot of excess air, the increased quantity of exhaust gas will lead to extra flue gas losses. Similarly, insufficient air for combustion results in wastage of fuel due to incomplete combustion and reduces the heat transfer efficiency due to soot buildup on heat transfer surfaces.

The amount of excess air required depends on the type of fuel and, in general, a minimum of about 10% to 15% excess air is required for complete combustion. This translates to about 2% to 3% of excess oxygen.

Boiler combustion efficiency, which indicates the ability of the combustion process to burn the fuel completely (with minimum excess air), can be measured by sampling the exhaust flue gas to find its composition and temperature using a combustion analyzer. Most good combustion analyzers are able to give a direct reading of the combustion efficiency based on the fuel used. If this facility is not available on the instrument used, a combustion efficiency versus oxygen (O_2) concentration chart (Fig. 5.8) can be used to estimate the combustion efficiency.

The drop in combustion efficiency due to excess air is dependent on the type of boiler and the amount of excess air. Based on the chart (Fig. 5.8), if the excess air is increased from 15% to 30%, the O_2 concentration will increase from 3% to 5% (as normal air contains 21% oxygen) and the resulting drop in efficiency will be about 1%.

For boilers operating at high-excess air levels, the combustion burner operation needs to be tuned to adjust the air-to-fuel ratio. This can normally be achieved by adjusting the mechanical linkages that control the flow of fuel and air to the burner to provide the correct ratio between the two at different operating loads for the boiler. Ideally, an oxygen (O_2) trim system should be installed which can continuously monitor the oxygen level in the flue gas and automatically adjust the air-to-fuel ratio to maximize combustion efficiency.

The amount of excess air also increases sometimes due to excessive draft created by the stack. If the stack is high, the natural draft created by the buoyancy of the combustion

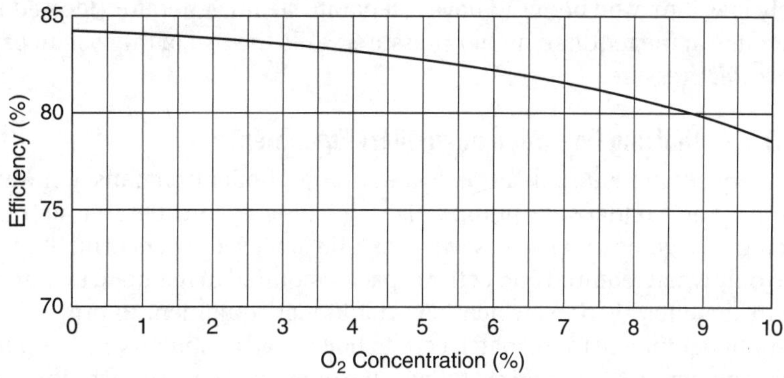

FIGURE 5.8   Combustion efficiency versus O_2 concentration (courtesy of Cleaver-Brooks).

gases can be significant. This effect can be overcome by having a draft-control system, which consists of an opening with a damper installed on the exhaust duct between the boilers and the stack, to automatically control the draft by opening or closing the damper.

The amount of excess air required for combustion also depends on the type of burner. Some old burners require much more excess air for complete combustion than others. Such burners can also be replaced with low excess-air burners to improve combustion efficiency.

5.5.2   Steam Pressure

Boilers have a maximum operating pressure rating, based on their construction, as well as a minimum value to prevent carryover of water. The actual operating pressure is normally set based on the requirements of the end users, while ensuring it is within the specified maximum and minimum values.

Since boiler efficiency depends on the operating pressure, if the operating pressure is set much higher than required, energy savings can be achieved by reducing it to match the actual requirements. Typically, reducing boiler pressure can help to improve boiler efficiency by 1% to 2%.

In addition to improving boiler efficiency, reducing steam pressure helps to reduce steam leaks and wastage due to overheating in some applications. Reducing pressure also lowers the temperature of the distribution piping, which helps to cut down on losses. Another benefit of reducing pressure is the reduction of flash steam from vents of condensate recovery systems.

The heat-carrying capacity (latent heat) of steam reduces with increase in pressure. Since many applications of steam involve condensing of steam in heat exchangers, it is recommended to keep the steam pressure at the lowest acceptable value to extract the maximum latent heat from steam.

However, it should be noted that when steam pressure is reduced, the distribution pipe sizing needs to be sufficient to transport the higher volume of steam.

If it is not possible to reduce the pressure of the entire system, parts of the distribution system can be operated at lower pressures by installing pressure-reducing valves at appropriate points in the distribution network.

In some systems, one steam user may require steam at a much higher pressure than the others. In such a system, if the steam usage of the high-pressure user is relatively low, it may be better to have a separate steam generator (located near the user) operating at the required higher pressure, while the rest of the system can be operated at a lower pressure.

5.5.3   Optimizing Operation of Auxiliary Equipment

Auxiliary equipment such as feedwater pumps, boiler draft fans, hot water circulating pumps, and condensate pumps also consume a considerable amount of energy. Therefore, significant energy savings can be achieved by ensuring that they are operated only when required and at the capacity required to maintain system requirements.

In some installations which have additional equipment to provide extra reliability (standby equipment) or to match certain boiler load conditions, plant operators may run more equipment than required to meet the operating load. In such situations, some auxiliary equipment can be switched off either manually or by using automatic controls.

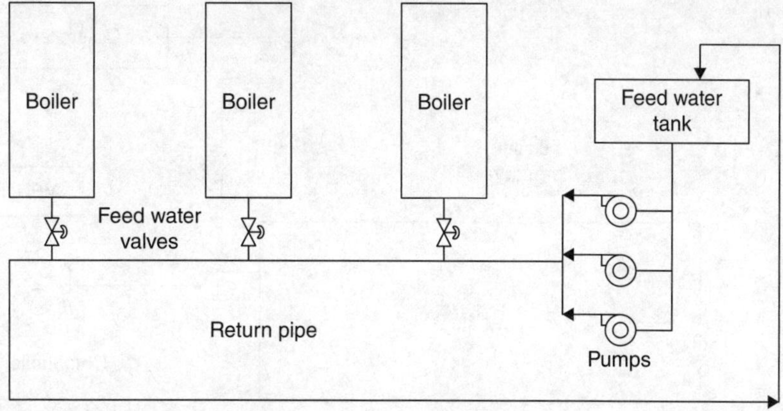

FIGURE 5.9   Feedwater pump arrangement for a multiple-boiler operation.

Generally, each boiler has its own feedwater pump which is automatically switched on and off to maintain the level of water in the boiler. Their operation is interlocked with the boiler so that the feedwater pump is switched off when the boiler is not in operation.

In larger systems, multiple boilers can be served by a common set of feedwater pumps, as shown in Fig. 5.9. In such an arrangement, the individual boilers take the required water flow by opening and closing the feedwater valves to maintain the water level in the boilers. The excess water is returned to the feedwater tank, which results in wastage of pumping energy.

This system can be improved to reduce the energy consumption of the pumps by varying the capacity (speed) of the feedwater pumps, which helps to maintain a set pressure in the feedwater header pipe, as shown in Fig. 5.10. A pressure-activated valve

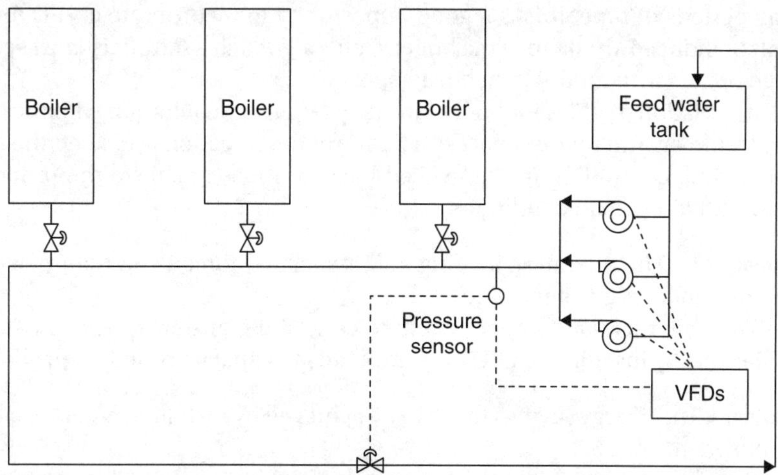

FIGURE 5.10   Suggested feedwater pump arrangement for a multiple-boiler operation.

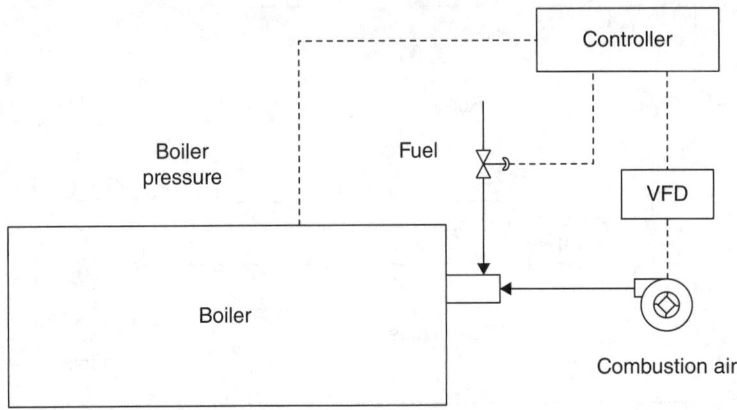

Figure 5.11   Application of a VFD for a boiler fan.

(normally closed) can be installed on the return pipe as a safety measure to open if the pressure exceeds a set value (which may occur due to failure of the pump control system).

Boiler fans used to create the draft necessary for combustion and carry the flue gases through the boiler normally operate at a constant speed and dampers are used to control the airflow to match boiler load conditions. In such systems, when the boiler operates at part load, a damper throttles the airflow by inducing a resistance across the path of the airflow. As a result, the energy consumption of the fan does not reduce proportionately to the airflow. However, if a variable-speed fan is used for this application (Fig. 5.11), due to the cube law [fan power $\propto$ (airflow rate)^3], the reduction in fan energy consumption would be proportional to the third power of the load. Therefore, theoretically, if the load on the boiler reduces by 20%, the energy consumption of the fan will be reduced by about 50% (0.8^3).

The application of this energy savings measure depends on the load profile of the boiler. If the load is highly variable and results in the boiler operating at low loads for long periods of time, this is a good opportunity to incorporate a VFD for the forced-draft or induced-draft fan of the boiler. Generally, such a retrofit is most economical for large boilers with modulating burners.

Installation of VFDs for boiler fans may require consultation with the boiler manufacturer to ensure that the necessary control modifications (to keep the damper fully open while controlling the fan speed based on load) achieve the proper air-to-fuel ratio at different load conditions.

Example 5.2   The operating loading and associated forced-draft fan power consumption of a boiler is given in Table 5.2.

The boiler uses a damper system to control the airflow rate. Estimate the savings achievable by installing a VFD to control the fan capacity based on the boiler load.

Solution   The energy savings that can be achieved by installing a VFD can be estimated as shown in Table 5.3.

Based on Table 5.3, the total savings is 296.8 kWh a day. This value can be multiplied by the number of operating days a year and the electricity tariff to calculate the annual cost savings.   ▲

Boiler loading	Operating hours a day	Fan motor power
100%	2	22
80%	4	21
60%	10	19
40%	8	16

TABLE 5.2   Boiler Operating Data for Example 5.2

Boiler loading	Operating hours a day	Fan motor power with damper	Fan motor power with VFD (kW)	Power savings (kW)	Energy savings (kWh)
100%	2	22	22	0	0
80%	4	21	11	10	40
60%	10	19	5	14	140
40%	8	16	1.4	14.6	116.8
Total					296.8

TABLE 5.3   Estimate of Savings for Example 5.2

5.5.4 Standby Losses

Standby losses take place when a boiler is not firing and the hot surfaces inside the boiler lose heat to colder air circulating inside it. Such air circulation can take place due to natural convection and purging.

Losses due to natural convection occur when the air in the boiler gets heated (by the hot surfaces), making it lighter and causing it to move up the stack circulating cold air through the boiler. This can be avoided if dampers are installed to prevent the circulation of air when the boiler is not being fired.

Purging losses take place when the boiler combustion space is purged by the fan before firing the burners to ensure that there is only air (to prevent possible explosions). Some burner systems also follow a purging cycle when firing stops. Losses due to purging can be reduced by minimizing the on–off cycle of the burner system. This can be achieved by using burners that have a high turndown ratio (ratio of maximum heat output to the minimum heat output of a burner) to enable the burner to function even at low loads without switching off the flame.

5.5.5 Minimizing Conduction and Radiation Losses

When a steam system is in operation, the surface temperature of boilers, auxiliary equipment, and distribution piping become much hotter than the surrounding areas. Therefore, they lose heat by radiation and conduction. The amount of heat lost depends on the surface temperature of the hot surface, which in turn depends on the insulation (thickness, thermal conductivity, and condition). To minimize heat loss, all hot surfaces should be insulated with material having sufficient resistance to heat transfer. Further, the insulation should be of adequate thickness and should be in a good condition.

For a typical boiler operating at full load, heat loss due to radiation and convection is about 0.5% to 1%. Since the radiation and convective losses remain the same, irrespective of boiler loading, a 1% loss at full load can increase to 4% when the boiler is operating at 25% load.

5.5.6 Heat Recovery from Flue Gas

A significant amount of heat energy is lost through flue gases as all the heat produced by the burning fuel cannot be transferred to the water or steam in the boiler. As the temperature of the flue gas leaving a boiler typically ranges from 150 to 250°C, about 10% to 20% of the heat energy is lost through it.

Therefore, recovering part of the heat from flue gas can help to improve the efficiency of the boiler. Heat can be recovered from the flue gas by passing it through a heat exchanger (commonly called an economizer) installed after the boiler, as shown in Fig. 5.12. The recovered heat can be used to preheat boiler feedwater, combustion air, or for other applications. The amount of heat recovered depends on the flue gas temperature and the temperature of the fluid to be heated.

One of the major problems associated with flue gas heat recovery is corrosion due to acid condensation. Acid condensation takes place when the flue gas is cooled below its acid dew point. The sulfur in the fuel combines with water to form sulfuric acid, which is corrosive. Therefore, the temperature of the flue gas needs to be maintained well above the acid dew point to prevent corrosion, unless a heat recovery system specially designed to withstand acid corrosion is used.

The acid dew point depends on the sulfur content of the fuel. Some typical values are given in Table 5.4.

The feasibility of installing a heat recovery system for flue gas depends on factors such as how much the stack temperature can be reduced, the inlet temperature of the fluid to be heated, and the operating hours of the boiler. Generally, the possible reduction in flue gas temperature should be at least 25°C to 30°C to make it economically viable to install a heat recovery system.

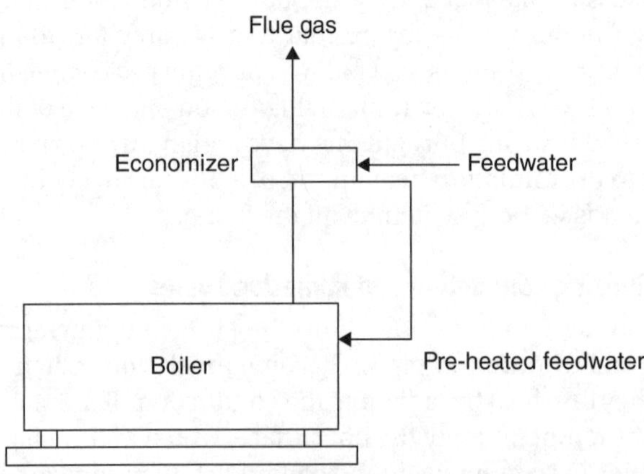

FIGURE 5.12 Arrangement of a typical economizer.

Fuel	Acid dew point temperature (°C)	Allowable exit stack temperature (°C)
Natural gas	66	120
Light oil	82	135
Low-sulfur oil	93	150
High-sulfur oil	110	160

TABLE 5.4   Acid Dew Point Temperature for Some Common Fuels

Since economizers induce extra pressure losses on the flue gas and the liquid being heated, care should be taken to ensure that the combustion fan and the pump for the liquid being heated have adequate capacity to overcome these losses.

Example 5.3   A diesel-fired boiler operates at a flue gas temperature of 220°C. Compute the energy savings achievable by installing an economizer to preheat feedwater. Assume that the air-to-fuel ratio is 15:1 and that the stack temperature has to be maintained above the acid dew point of 150°C. The monthly diesel consumption is 330,000 L. The density of diesel is 900 kg/m³ and the specific heat capacity of flue gas is 1.1 kJ/kg K.

Solution

$$\text{Diesel consumption rate} = 333,000 \text{ L/month} = 333 \text{ m}^3/\text{month}$$
$$= 333 \times 900 \text{ kg/month}$$
$$= (333 \times 900)/(24 \times 30) = 416.25 \text{ kg/h}$$

$$\text{Air fuel ratio} = 15:1$$
$$\text{Airflow rate} = 15 \times 416.25 = 6243.75 \text{ kg/h}$$
$$\text{Total mass flow rate of flue gas} = 416.25 + 6243.75 = 6660 \text{ kg/h} = 6660/3600$$
$$= 1.85 \text{ kg/s}$$
$$\text{Heat recovered, } Q = m \times C_p \times \Delta T$$

where m = mass flow rate of flue gas, C_p = specific heat capacity of flue gas, and ΔT = reduction in temperature of flue gas.

$$\text{Hence, } Q = 1.85 \times 1.1 \times (220 - 150) = 142.45 \text{ kW} \quad \blacktriangle$$

5.5.7   Flash Steam Recovery

The temperature of water at high pressure will be at a temperature higher than the boiling point of water at atmospheric pressure. As can be seen from the sample steam table (Table 5.1), water, for instance, at 6 bar will be at a temperature of 158.8°C, which is much higher than the boiling point of 100°C at atmospheric pressure. Therefore, when this high-pressure water is released to atmospheric pressure at a steam trap or as blowdown (explained later), some of the water will flash back into steam. This flash steam, if produced in large quantities, can be used as low-pressure steam or condensed back into feedwater.

Example 5.4   An amount of 0.25 kg/s of condensate at 8 bar (absolute) is released through a steam trap. Compute the percentage and amount of flash steam produced.

Solution

$$\text{Enthalpy of water at 8 bar} = 721 \text{ kJ/kg (from Table 5.1)}$$
$$\text{Enthalpy of water at atmospheric pressure} = 419 \text{ kJ/kg}$$
$$\text{Excess energy} = 721 - 419 = 302 \text{ kJ/kg}$$
$$\text{Enthalpy of evaporation at atmospheric pressure} = 2257 \text{ kJ/kg}$$
$$\text{Percentage of flash steam produced} = (\text{excess energy/enthalpy of}$$
$$\text{evaporation}) \times 100\%$$
$$= (302/2257) \times 100\%$$
$$= 13\%$$
$$\text{Amount of flash steam produced} = 0.25 \times 0.13 = 0.0325 \text{ kg/s} \quad \blacktriangle$$

5.5.8   Automatic Blowdown Control

Boiler blowdown is part of the water-treatment process and involves removal of sludge and solids from the boiler. Make-up water used for boilers contains various impurities. As water is converted into steam, the concentration of the impurities that remain in the boiler increases. If this concentration is allowed to increase, it will lead to accelerated corrosion, scaling, and fouling of the heat-transfer surfaces of the boiler. Therefore, it is necessary to remove part of the concentrated water from the boiler and replace it with fresh water.

The quality of water in the boiler is normally based on the TDS (total dissolved solids) level measured in ppm (parts per million). For typical two-pass and three-pass fire tube boilers, the allowable TDS level is 3000 to 3500 ppm, while for water tube boilers, it is about 1500 ppm.

The blowdown rate depends on the rate of steam production, TDS of the feedwater, and the allowable maximum TDS and can be expressed as follows:

$$\text{Blowdown rate (kg/h)} = (S \times F_T)/(M_T - F_T)$$

where S = steam production rate (kg/h), F_T = feedwater TDS (ppm), and M_T = maximum allowable TDS (ppm).

Example 5.5   The TDS level of the boiler feedwater is 200 ppm. Estimate the blowdown rate for an 8000-kg/h boiler if the allowable maximum TDS is 3000 ppm.

Solution

$$\text{Blowdown rate (kg/h)} = (S \times F_T)/(M_T - F_T)$$
$$= (8000 \times 200)/(3000 - 200)$$
$$= 571.4 \text{ kg/h} \quad \blacktriangle$$

Boiler blowdown can be intermittent, where a fixed quantity of water is drained periodically, or continuous, where a small amount of water is removed continuously to maintain the quality of water within acceptable limits.

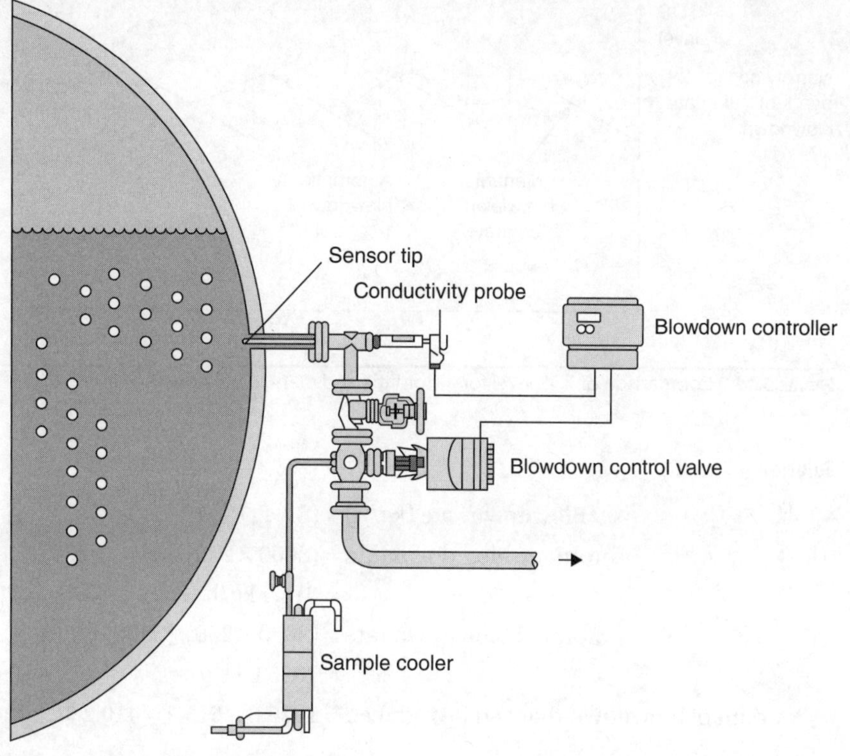

FIGURE 5.13   Typical arrangement of automatic blowdown control (courtesy of Spirax Sarco).

Blowdown involves discharge of water at steam temperature, which has to be replaced by an equivalent amount of cold water. Energy losses resulting from blowdown can be minimized by installing automatic blowdown systems to reduce the amount of blowdown and recovering heat from blowdown (Fig. 5.13).

Automatic blowdown control systems monitor the pH and conductivity of the boiler water and allow blowdown only when required to maintain an acceptable level of water quality. Automatic blowdown systems are preferred as they can maintain the TDS level close to the maximum allowable value. Manual blowdown (normally every 8 h), on the other hand, requires the TDS level to be reduced to a much lower value during the actual blowdown to ensure the TDS level does not exceed the maximum allowable value until the next blowdown cycle, resulting in a much lower average TDS level. A comparison of TDS variation for manual and automatic blowdown control is shown in Fig. 5.14.

Example 5.6   An 8000-kg/h boiler uses manual blowdown. The average TDS is maintained at 2200 ppm. Compute the reduction in blowdown rate that can be achieved by installing an automatic TDS control system to maintain the average TDS level at 3500 ppm. The TDS of the feedwater is 250 ppm.

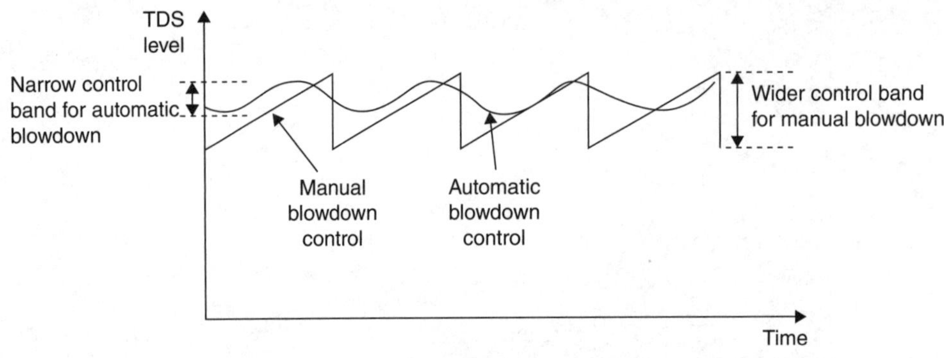

Figure 5.14   Comparison of TDS level for automatic and manual blowdown.

Solution

$$\text{Blowdown rate (kg/h)} = (S \times F_T)/(M_T - F_T)$$
$$\text{Automatic blowdown rate} = (8000 \times 250)/(3500 - 250)$$
$$= 615.4 \text{ kg/h}$$
$$\text{Manual blowdown rate} = (8000 \times 250)/(2200 - 250)$$
$$= 1025.6 \text{ kg/h}$$
$$\text{Amount of blowdown that can be reduced} = 1025.6 - 615.4 = 410.2 \text{ kg/h} \quad \blacktriangle$$

5.5.9   Heat Recovery from Blowdown

Heat recovery from blowdown involves using a heat exchanger to preheat cold make-up water using the blowdown. Such systems are feasible for boilers that operate most of the year using at least 5% of make-up water.

Flash steam (explained earlier) also can be recovered from blowdown and a typical arrangement is shown in Fig. 5.15.

Example 5.7   The TDS level of a 10,000-kg/h boiler is maintained at 2,500 ppm using an automatic TDS control system. The TDS of the feedwater is 200 ppm and the boiler operating pressure is 10 bar (absolute). Compute the amount of heat energy in the blowdown and the amount of flash steam that can be recovered from the blowdown.

Solution

$$\text{Blowdown rate (kg/h)} = (S \times F_T)/(M_T - F_T)$$
$$= (10,000 \times 200)/(2,500 - 200)$$
$$= 869.6 \text{ kg/h}$$
$$= 0.24 \text{ kg/s}$$
$$\text{Enthalpy of blowdown at 8 bar, } h_f = 763 \text{ kJ/kg (Table 5.1)}$$
$$\text{Heat energy in the blowdown} = 0.24 \text{ (kg/s)} \times 763 \text{ (kJ/kg)} = 183 \text{ kW}$$
$$\text{Enthalpy of water at atmospheric pressure} = 419 \text{ kJ/kg}$$
$$\text{Excess energy} = 763 - 419 = 344 \text{ kJ/kg}$$

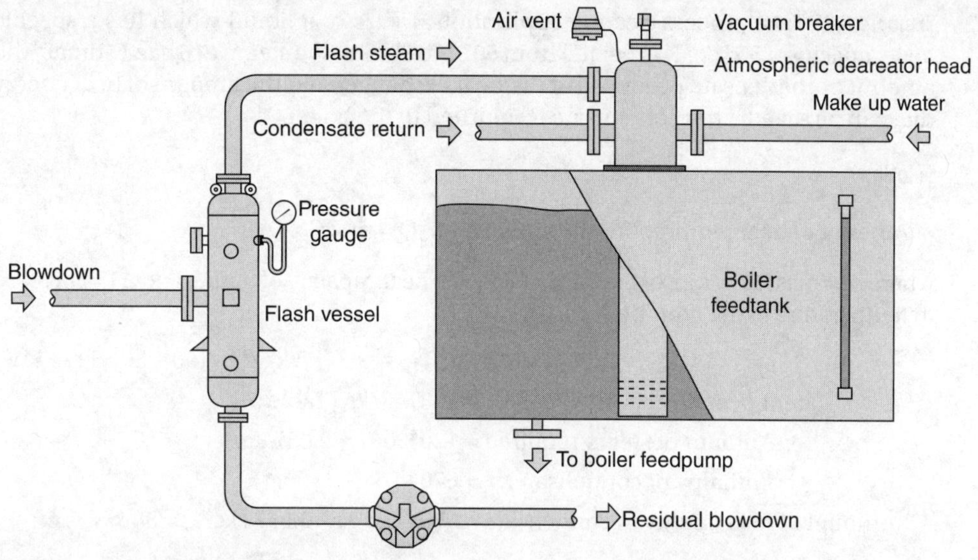

Figure 5.15 Typical blowdown flash steam recovery system (courtesy of Spirax Sarco).

$$\text{Enthalpy of evaporation at atmospheric pressure} = 2257 \text{ kJ/kg}$$

$$\text{Percentage of flash steam} = (\text{excess energy/enthalpy of evaporation}) \times 100\%$$

$$= (344/2257) \times 100\%$$

$$= 15\%$$

$$\text{Amount of flash steam} = 0.24 \times 0.15 = 0.036 \text{ kg/s} \quad \blacktriangle$$

5.5.10 Condensate Recovery

In most steam systems, steam is used mainly for heating by extracting its latent heat (h_{fg}). The resulting condensate is at steam temperature and still contains a considerable amount of heat energy (h_f). From Table 5.1, if steam is used at 8 bar (absolute), then the condensate enthalpy (h_f) will be 721 kJ/kg, which is 26% of the total enthalpy of steam (2769 kJ/kg) and will be lost if condensate is not returned to the system. Therefore, returning condensate to the boiler feedwater tank will result in significant fuel energy savings.

Since condensate is distilled water, it is ideal for use as boiler feedwater. Therefore, condensate recovery helps to reduce water consumption (water cost), water treatment cost, and blowdown.

Usually, a low feedwater temperature or high make-up water flow indicates that less condensate is recovered. If the make-up water flow is metered, in applications which do not consume live steam (such as open sparge coils and direct steam injection systems), the difference between the amount of steam produced and make-up water flow will give an indication of the amount of condensate that is not recovered.

Example 5.8   In a process heating application, 1 kg/s of a liquid which has a specific heat capacity of 3 kJ/kg K is heated from 50°C to 90°C using steam at 6 bar. Estimate the amount of condensate produced at the heat exchanger and the amount of heat energy that can be saved if the condensate is returned to the boiler.

Solution

Amount of heat required for the application, $Q = m \times C_p \times \Delta T$

where m = mass flow rate of liquid, C_p = specific heat capacity of liquid, and ΔT = increase in temperature of the liquid.

$$Q = 1 \; (\text{kg/s}) \times 3 \; (\text{kJ/kg K}) \times (90 - 50) = 120 \; \text{kW}$$

Latent heat of steam at 6 bar, $h_{fg} = 2087$ kJ/kg (Table 5.1)

Amount of steam required = $120/2087 = 0.057$ kg/s

Enthalpy of condensate, $h_f = 670$ kJ/kg

Amount of heat energy in condensate = $670 \; (\text{kJ/kg}) \times 0.057 \; (\text{kg/s}) = 38 \; \text{kW}$ ▲

5.5.11   Steam Traps

Steam traps are used in steam systems to remove condensate and noncondensable gases. They are mainly used for steam heating coils and for condensate removal from steam headers.

Steam traps are generally classified as thermostatic, mechanical, or thermodynamic (Fig. 5.16). Thermostatic steam traps are designed to work based on the difference in temperature between steam and condensate. They contain a bimetallic strip or bellows to allow subcooled condensate to be removed while preventing live steam, which is at a higher temperature, from passing through. While bellows-type thermostatic traps can be used for steady light loads on low-pressure systems, bimetallic traps can be used for jacketed piping, steam tracers, and heat-transfer equipment, which can accommodate backup of condensate.

Bucket and float-type traps are common types of mechanical steam traps. As the names imply, they have floating balls or buckets that operate on the buoyancy of condensate to mechanically open and close ports in the traps to discharge only condensate. They usually have built-in air-venting features and are used for continuous and intermittent loads. They are commonly used on steam heat exchanger coils.

Thermodynamic traps operate based on the difference in flow characteristics of steam and condensate. When air or condensate enters the trap, a disc lifts up to allow it to be discharged. When steam enters the trap, due to its increased velocity (higher velocity pressure), the static pressure below the disc is reduced, which lowers the disc, closing the trap. They are often used for condensate removal from main steam distribution pipes.

The operation of steam traps is important because if they fail to operate properly and allow live steam to pass through them from the steam side to the condensate side, they result in obvious loss of energy. In addition, if the traps are unable to remove air at start-up times or if they are unable to remove condensate at a sufficient rate, the resulting reduced capacity and longer periods to heat up would also result in energy wastage.

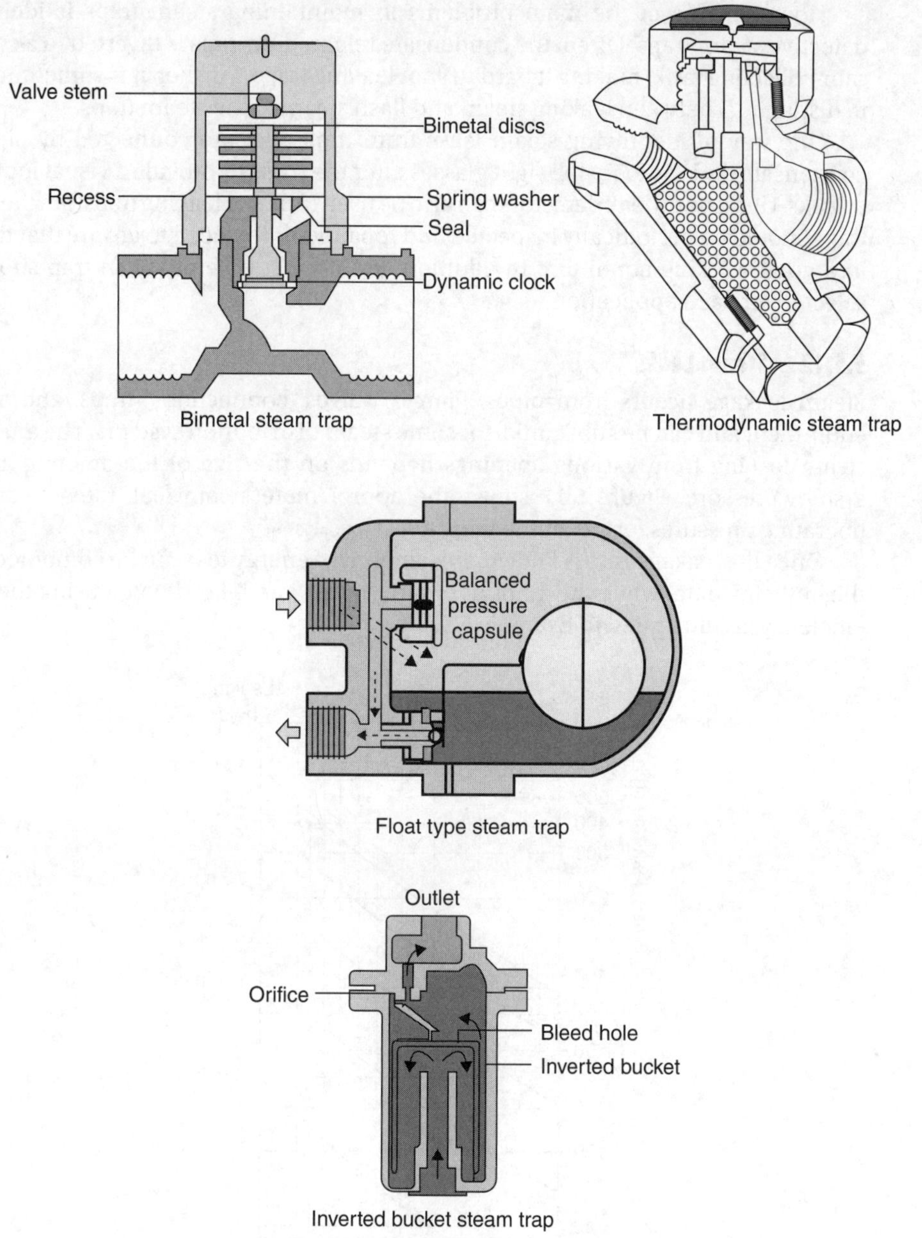

Figure 5.16 Common types of steam traps (courtesy of Spirax Sarco).

Over time, internal parts of steam traps wear out and result in failure to open and close properly. While an open trap would result in loss of live steam, a closed trap could result in loss of heat-transfer area and water hammering. Water hammering can eventually result in damage to valves and other components in steam systems, which could result in steam leaks.

However, one of the main problems in maintaining steam traps is identifying defective steam traps. Often, the condensate released by traps is diverted to a condensate-collecting tank, making it hard to spot leaking traps. Further, it is sometimes hard to distinguish between leaking steam and flash steam at the steam traps.

One way of identifying steam leaks from traps that are connected by piping to condensate tanks is to install sight glasses after the traps to provide a visual indication of leaks. Ultrasound leak detectors can also be used to detect leaking traps. Furthermore, traps should be periodically inspected and repaired or replaced to ensure that they are in a good working condition. In addition, the correct type of steam trap should be selected for each application.

5.5.12   Steam Leaks

Steam leakage occurs from pipes, flanges, valves, connections, traps, and process equipment and can be substantial for some steam distribution systems. The amount of steam leaking from various openings depends on the size of the opening and the system pressure. Figure 5.17 shows the approximate steam leak rates at different operating pressures for various leaking hole sizes.

Once the leakage rate is known, the amount of energy lost can be computed using the enthalpy data, while the amount of fuel wasted can be estimated using the boiler efficiency, as illustrated in Example 5.9.

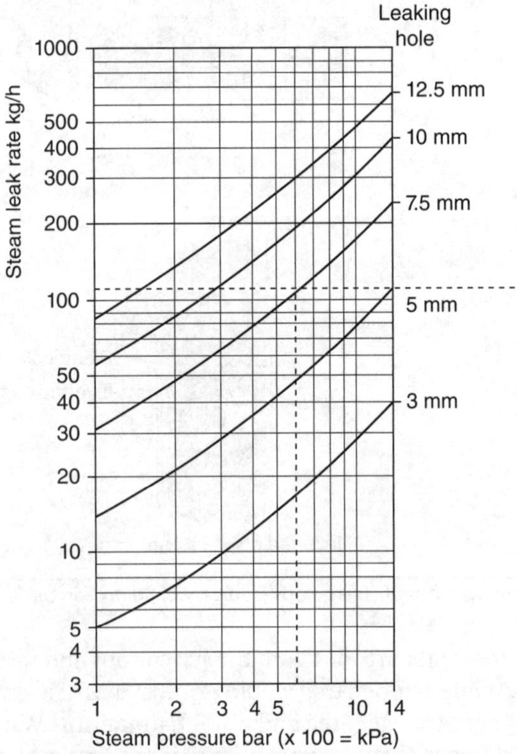

Figure 5.17   Steam losses through leaks (courtesy of Spirax Sarco).

Example 5.9   The total steam leak rate for a plant is estimated to be 100 kg/h. If the steam pressure is 7 bar (absolute), compute the amount of energy wasted due to the steam leaks. The boiler uses natural gas as the fuel, which has a calorific value of 55,000 kJ/kg and the operating boiler efficiency is 88%. Estimate the reduction in fuel use by the boiler that can be achieved if all the steam leaks in the plant are rectified.

Solution

$$\text{Enthalpy of steam at 7 bar, } h_g = 2764 \text{ kJ/kg (Table 5.1)}$$
$$\text{Total steam leak rate} = 100 \text{ kg/h} = 0.028 \text{ kg/s}$$
$$\text{Amount of heat energy wasted} = 0.028 \text{ (kg/s)} \times 2764 \text{ (kJ/kg)} = 77.4 \text{ kW}$$
$$\text{Amount of input heat energy wasted} = 77.4/0.88 = 87.95 \text{ kW}$$
$$\text{Amount of fuel wasted} = 87.95 \text{ (kW)}/55,000 \text{ (kJ/kg)}$$
$$= 0.0016 \text{ kg/s} = 5.76 \text{ kg/h} \quad \blacktriangle$$

5.5.13   Feedwater Tank

The feedwater tank is a very important part of any steam system. It provides a reservoir of returned condensate and fresh make-up water for the boilers. The feedwater tank gives a good indication of the system performance. Excessive feedwater temperature may indicate that some traps may be passing live steam, while a high make-up water flow may indicate that some condensate is not being returned to the tank.

Feedwater is normally hot due to returned condensate and recovery of heat from other sources. Therefore, the tank should be elevated to prevent hot water being flashed off as steam at the feedwater pump inlet, in order to prevent cavitation.

Since the feedwater tank is hot, steps should be taken to minimize heat losses from the tank. Other than insulating the tank, since a great amount of losses usually take place at the water surface, the top of the tank should be covered.

5.5.14   Fouling and Scaling in Boilers

Fouling, scaling, and soot build-up on heat-transfer surfaces of boilers acts as an insulator and leads to reduced heat transfer. This results in lower heat transfer to water in the boiler and higher flue gas temperature. If at the same load conditions and same excess air setting the flue gas temperature increases with time, this is a good indication of increased resistance to heat transfer in the boiler. Data for energy wastage due to scaling are shown in Table 5.5.

Scale thickness (mm)	Energy wastage (%)
0.4	1
0.8	2
1.2	3
1.6	4

TABLE 5.5   Estimated Energy Wastage Due to Scaling

When this occurs, the boiler heat-transfer surface should be cleaned. On the fire side, surfaces should be cleaned of soot, while on the water side, scaling and fouling should be removed. For boilers using gas and light oil, it is generally sufficient to clean the fire-side surfaces once a year. However, for boilers using heavy oil, cleaning may need to be done several times a year.

In addition, preventive steps should also be taken. For scaling, as it is caused by inadequate water treatment, steps should be taken to improve water softening and maintaining a lower TDS level. For soot buildup, which is normally due to a defective burner or insufficient air for combustion, steps should be taken to repair or retune the combustion system.

CHAPTER 6

Process Cooling Systems

6.1 Introduction

Many manufacturing processes require heat input to change the temperature or state of raw materials. In addition, industrial plants can also generate heat as a result of chemical reactions or due to frictional or other mechanical losses.

In most cases, the heat generated or added needs to be removed to maintain a desired product temperature or for safety reasons. Heat is commonly removed using cooling systems with a fluid such as water or air, which is circulated through heat exchangers or jackets surrounding vessels.

Industrial cooling systems can be once-through systems where ambient air or water is passed through heat exchangers to absorb heat after which it is returned to the same source.

Other types of self-contained cooling systems use a fluid circulating in a closed loop to remove heat from the process but contain another system to remove the heat from this fluid. The design of such closed-loop systems depends on the required temperature of the cooling medium and can be generally categorized as follows:

- Cooling tower systems
- Chilled-water and low-temperature cooling systems
- Refrigeration systems

This chapter contains a description of the commonly used industrial cooling systems and how they can be optimized to minimize energy consumption.

6.2 Once-Through Systems

Once-through air-cooled systems use ambient air as the cooling medium to remove heat from the material being cooled through a heat exchanger.

Similarly, once-through water-cooled systems use water from a heat sink such as a river, lake, or sea. A typical water-cooled system is shown in Fig. 6.1, where water taken from the heat sink is pumped through one or more heat exchangers to absorb heat. Thereafter, the water is discharged back into the same source. Although such systems do not consume energy for cooling, they can have an adverse impact on the environment due to localized heating of the water source and possible contamination due to leakage.

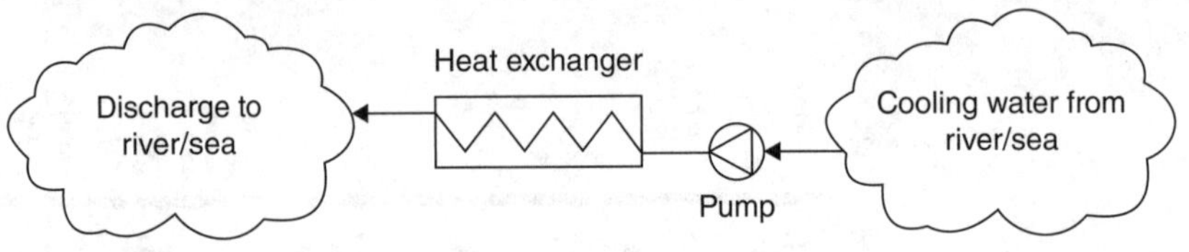

FIGURE 6.1   Arrangement of a typical once-through system.

The main energy consumers in a once-through system are the pumps and fans, which are used to circulate water or air through the cooling system. Therefore, energy consumption in such a system can be minimized by optimizing the pumping or fan system. Such optimizing measures are explained in detail in Chaps. 3 and 4 and include reducing pressure losses in the system (piping, filters, fittings, heat exchangers, etc.), reducing the flow rate to the minimum value required to achieve the design system temperature, and selecting the most efficient pumps or fans for the application.

6.3   Cooling Tower Systems

Many industrial processes that require a cooling medium at a temperature close to the ambient temperature use cooling tower water systems. A typical system consists of cooling towers and pumps that circulate water through heat exchangers to absorb heat, as shown in Fig. 6.2. Such systems reject the heat from the circulating water through evaporation and heat conduction between the warm water and the ambient air in the cooling towers and can achieve a water temperature of a few degrees above the wet-bulb temperature of ambient air.

Cooling towers are heat rejection devices that are designed to transfer heat to the atmosphere. They mainly consist of a water spray system that spreads the warm water being cooled over a fill packing that acts as a heat-transfer medium by increasing the contact surface area. Natural drafts or forced drafts using fans are used to induce

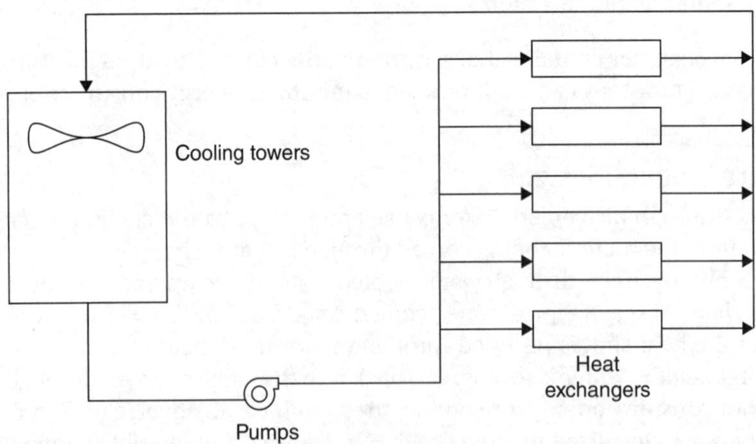

FIGURE 6.2   Typical cooling tower water system.

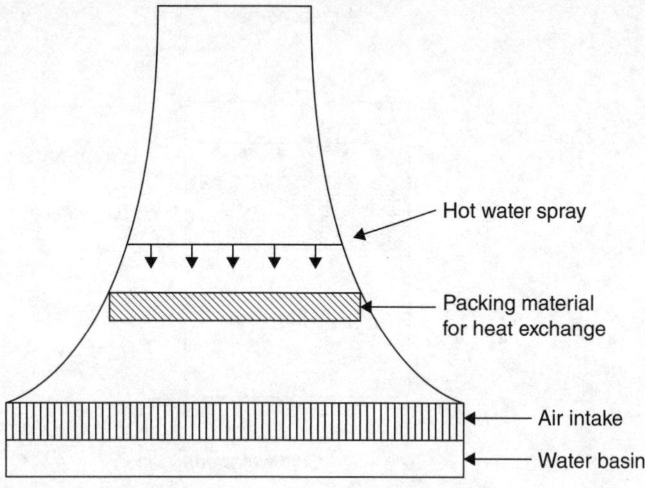

Figure 6.3 Typical natural-draft cooling tower.

ambient airflow through the cooling towers to facilitate heat transfer between the warm water and ambient air.

Cooling towers reject heat mainly by evaporative cooling. When water is sprayed in cooling towers, some of the water evaporates absorbing heat from the surrounding water, thereby cooling it. The amount of latent heat transferred depends on the moisture content of the air; the more dry the air (lower the wet-bulb temperature), the more will be the transfer of latent heat. In addition, sensible cooling also takes place between the warmer water and colder air. The amount of sensible cooling depends on the dry-bulb temperature of air. Therefore, the amount of heat rejected from cooling towers depends on both dry-bulb and wet-bulb temperatures of the outdoor air.

Cooling towers can be classified as natural draft or mechanical draft. Natural-draft cooling towers (Fig. 6.3) use a tall chimney with a hyperbolic shape to induce the warm moist air to rise by buoyancy, thereby creating a natural circulation of the air. Such cooling towers do not require fans and therefore do not use energy. However, they require very large concrete structures and are generally used for very large heat rejection capacities such as in power generation.

Mechanical-draft cooling towers use fans to induce airflow through cooling towers. The airflow in cooling towers can be either induced draft, where fans pull air out of cooling towers, or forced draft, where fans force air into cooling towers. In addition, the relative direction of air and warm water flows can be counterflow or crossflow.

Therefore, cooling towers can be forced-draft counterflow (Fig. 6.4), induced-draft counterflow (Fig. 6.5), forced-draft crossflow (Fig. 6.6), or induced-draft crossflow (Fig. 6.7).

6.3.1   Energy-Saving Measures for Cooling Tower Systems

Energy savings can be achieved for cooling tower systems by optimizing the circulating pumps and the cooling tower fans as described below.

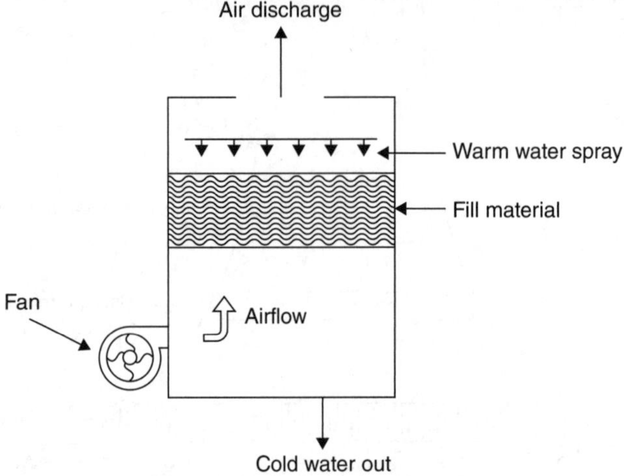

Figure 6.4 Arrangement of a forced-draft counterflow cooling tower.

Optimizing Pumping Systems

Energy consumed by the pumps can be reduced by minimizing losses in the piping system, selecting pumps to operate at the highest efficiency and reducing the circulating flow rate to meet the minimum required for the cooling process. These energy-saving measures are also applicable to all pumping systems and are described in detail in Chap. 3.

In most plants, cooling tower water cooling systems serve numerous heat exchangers serving various processes, as shown in Fig. 6.8. Often, these processes are operated in batches and require cooling only at specific times during each cycle. However, in

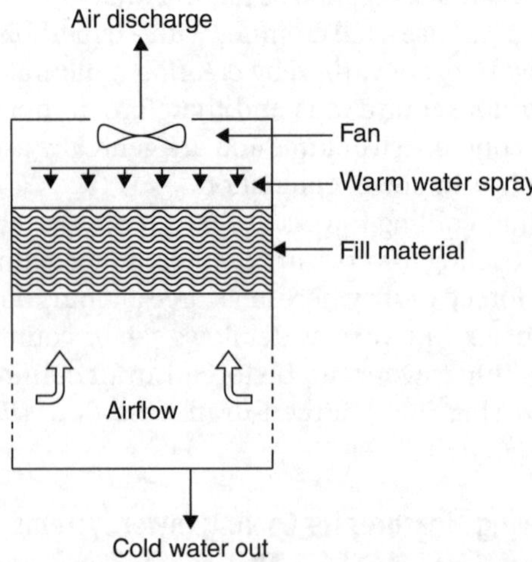

Figure 6.5 Arrangement of an induced-draft counterflow cooling tower.

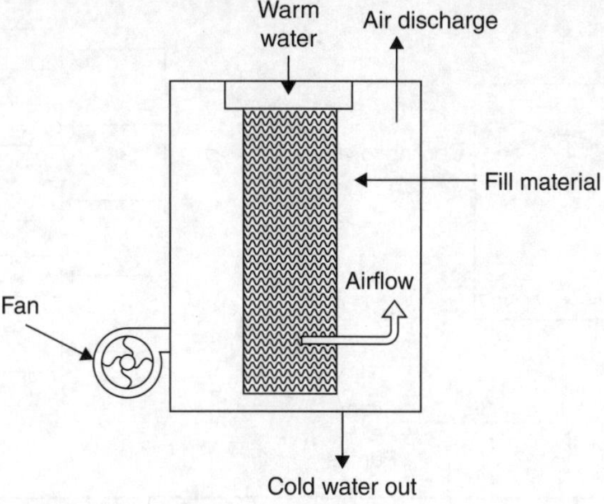

FIGURE 6.6   Arrangement of a forced-draft crossflow cooling tower.

most cases, the cooling water is allowed to flow through the heat exchanger even when heat removal is not necessary, wasting pumping energy.

In such situations, motorized valves can be installed to shut off the flow to the respective heat exchanger when cooling is not required and the circulating pumps fitted with variable-speed drives and a control system to maintain a minimum set pressure (P) in the circulation system, as shown in Fig. 6.9. In addition, temperature controls can be installed to further minimize the circulating flow rate where the valves automatically adjust the water flow rate through each heat exchanger to allow only the minimum flow rate required to maintain the desired cooling rate.

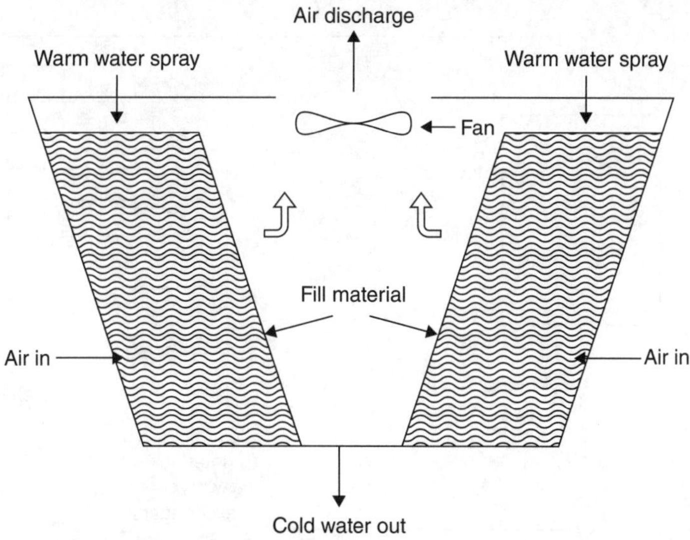

FIGURE 6.7   Arrangement of an induced-draft crossflow cooling tower.

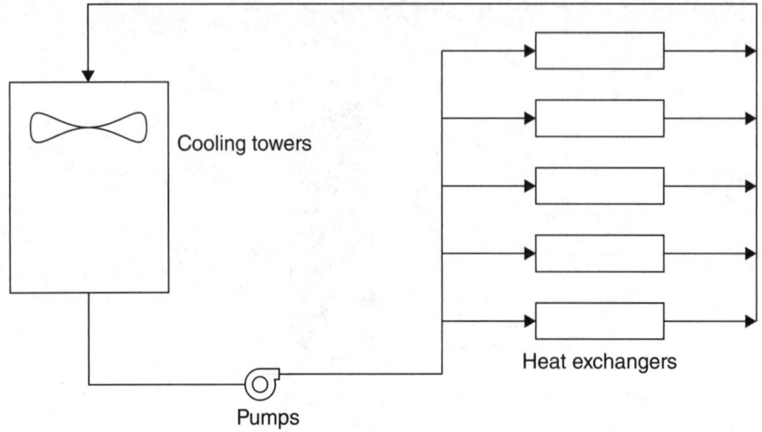

FIGURE 6.8   Arrangement of a cooling tower water cooling system serving heat exchangers.

Optimizing Cooling Towers

Energy consumed by cooling tower fans can be minimized by sizing cooling towers to provide the required capacity, matching the operating capacity of cooling towers to match load requirements, using high-efficiency fans, and following best practices in the installation and maintenance of cooling towers.

Capacity Control

Capacity of cooling towers is dependent on the airflow through them. Therefore, when the heat rejection load is lower than the design value (such as when only part of the plant is in operation), it is not necessary to run the cooling towers at full capacity.

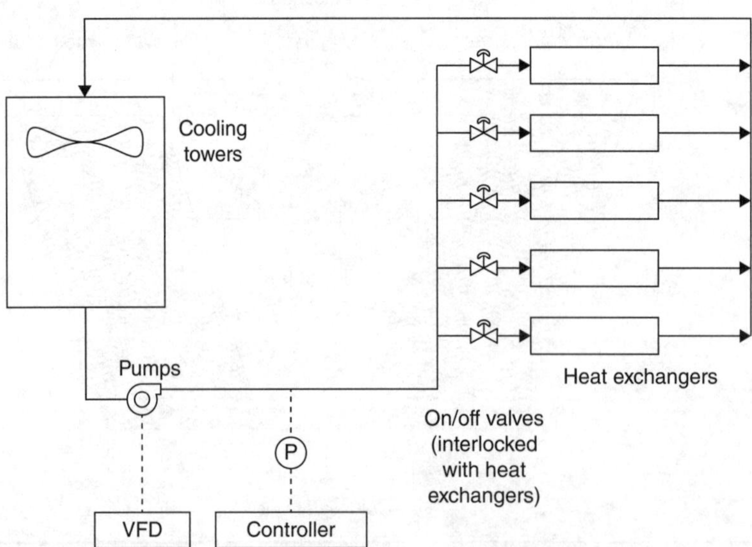

FIGURE 6.9   Arrangement of a variable-flow water circulation system.

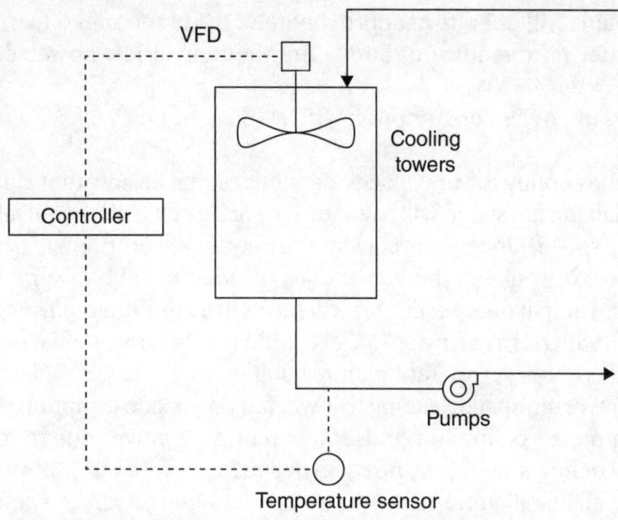

Figure 6.10   Cooling towers with variable-speed fans.

The cooling tower capacity can be reduced by reducing the airflow, which would result in lower fan energy consumption.

One way of achieving this is by cooling tower fan cycling where some fans are switched on/off to control the supply water temperature. This, however, can result in a swing in water temperature and can cause premature wear and tear of the cooling tower fan motor drives.

A better way is to use variable-frequency drives (VFDs) to control the cooling tower fan speed. As shown in Fig. 6.10, the speed of the cooling tower fans in operation can be modulated to maintain a set supply temperature. The easiest control strategy is to maintain the water supply temperature at the design value. Alternatively, the set-point can be adjusted to maintain a cooling tower approach temperature (difference between the water supply and wet-bulb temperatures) so that the water supply temperature is always a few degrees higher than the wet-bulb temperature.

Since theoretically the power consumed by fans is proportional to the cube of the fan speed, when, for instance, the cooling load drops to 80%, the speed of cooling tower fans can also be reduced accordingly, resulting in a drop in power consumption of about 50% ($0.8^3 = 0.51$). Therefore, this control strategy can lead to very significant savings from cooling tower fans at part load. Use of VFDs to control the cooling tower capacity rather than fan staging also leads to reduction in wear and tear on the drives due to lower fan speed and less drift losses (water losses) due to lower air velocity.

Example 6.1   A cooling tower that has two 15-kW fans (constant speed) is designed to cool water from 45 to 30°C (15°C range).

If under normal operating conditions the temperature of the warm water entering the cooling tower is 38°C, compute the fan power savings that can be achieved by installing a VFD to maintain the same supply water temperature.

Solution   The load on the cooling tower is only 53% (8°C operating temperature difference divided by the design value of 15°C). Therefore, if a VFD is installed on this cooling tower, the fans will operate at approximately 53% of the speed to maintain the required leaving water temperature of 30°C. The theoretical fan power consumption will be $(0.53)^3 \times 15 \times 2 = 4.5$ kW.

Savings in power consumption will be $(30 - 4.5) = 25.5$ kW.   ▲

Since the cooling tower capacity depends on the airflow rate, operating two identical towers at half the fan speed will result in a capacity that will be equivalent to operating one tower at full speed. Since, theoretically, the power consumption of the fans is proportional to the cube of the speed, the power consumption of each tower fan will drop to 12.5% $(0.5^3 = 0.125)$ at half the speed. This will result in a combined power consumption of 25% $(12.5\% \times 2)$, which is a saving of 75% of the fan power consumed if only one cooling tower is operated. Therefore, operating more cooling towers in parallel to meet the same load will allow lower fan speed, leading to lower fan power consumption. An additional benefit of running more cooling towers is lower pumping power due to lower pressure losses across the cooling tower spray nozzles (due to lower water flow through each tower).

Almost all installations have one or more cooling towers as standbys to allow servicing and repairing of running cooling towers. Therefore, if the standby towers are run in parallel with the duty cooling towers and if all cooling tower fans are fitted with VFDs, they can be run at a lower speed to produce the same amount of heat dissipation.

Example 6.2   A particular installation has three cooling towers (each 15 kW) where only two are operated and the remaining one acts as a standby tower. Compute the savings achievable if all three cooling towers are operated with VFDs to maintain the same supply water temperature.

Solution   If all three cooling tower fans are operated at 2/3 capacity each (three towers running at 2/3 capacity will provide the equivalent of two towers), savings in fan power can be estimated as follows:

Fan power consumption of each tower will be $15 \times (2/3)^3 = 4.4$ kW (about 30% of full-load power)

Total power consumption of all three fans will be $4.4 \times 3 = 13.2$ kW

Therefore, total savings will be $(15 \times 2) - 13.2 = 16.8$ kW

The above savings estimation is based on the system operating at the design capacity of two cooling towers. Further savings will also be achieved when the heat rejection load is low.   ▲

High-Efficiency Fans

Fans installed in cooling towers are designed to provide the required air circulation rate to achieve the desired heat rejection rate. The energy consumed by cooling tower fans depends on the airflow rate, pressure losses in the system, and the efficiency of the fan (Chap. 4).

Reducing the airflow rate in cooling towers to reduce fan energy consumption was discussed in the preceding section, while pressure losses in cooling towers are dependent on their design and, hence, cannot be changed by the end users. Therefore, the other opportunity for reducing fan energy consumption of a cooling tower is to improve the efficiency of the fan.

Figure 6.11   FRP blade with an aerodynamic profile (courtesy of Impact Group India).

The typical efficiency of axial-flow fans used for cooling towers is about 60%. The main reason for this relatively low efficiency is because metallic blades that are commonly used for cooling tower fans are manufactured by extrusion or casting processes where it is not possible to produce ideal aerodynamic blade profiles.

However, cooling tower fan blades can also be made from fiber reinforced plastic (FRP), which are normally hand molded and therefore easier to produce with optimum aerodynamic profiles. Such FRP fan blades are designed to have twisted and tapered blade profiles and can achieve an efficiency of about 80%.

An FRP fan blade with an aerodynamic profile is shown in Fig. 6.11. Cooling tower fan retrofit projects that replace the conventional metallic fan blades with FRP blades can result in fan energy savings of 20% to 30%.

Installation of Cooling Towers

Since the performance of cooling towers depends on the airflow through them, cooling towers should be installed such that air can flow freely into them. As shown in Fig. 6.12, cooling towers should be sufficiently spaced so that the air intakes of cooling towers are not too close to each other. Furthermore, cooling tower air intakes should not be less than a minimum distance from obstructions such as walls. The minimum distance to be maintained between cooling towers and from obstructions is normally specified by the cooling tower suppliers based on the design.

Care should also be taken to ensure that the warm and moist air being discharged from cooling towers is not recirculated back into the air intakes, as it will lead to a drop in tower performance. In some situations, an extension duct can be fitted to the cooling tower discharge to direct the airflow away from the air intakes, as shown in Fig. 6.13.

Maintenance of Cooling Towers

The ability of a cooling tower to provide its design cooling capacity depends on the operation of the water spray system, the fill, and the fan. Therefore, regular maintenance should ensure that the water spray system is able to properly spread the

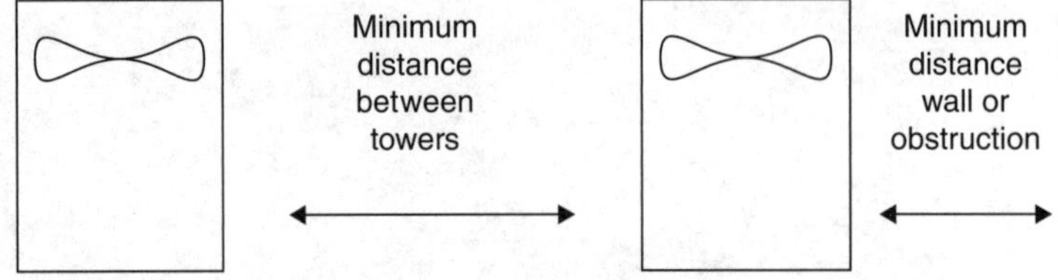

FIGURE 6.12 Spacing of cooling towers.

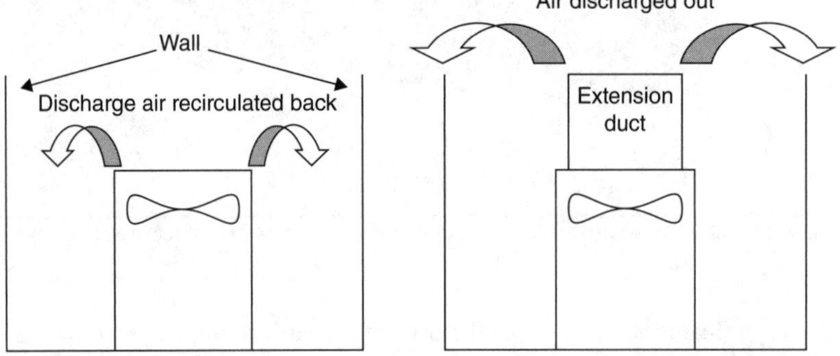

FIGURE 6.13 Extension duct used to prevent recirculation.

water flow, the fill is in good condition, and the fan is able to operate at the required speed.

If the spray system is defective, warm water may flow directly into the discharge basin of the cooling tower, rather than being sprayed onto the fill material, which is meant to facilitate the heat transfer between the warm water and the ambient air. Similarly, if the infill is damaged or blocked, water will also flow directly into the discharge basin, rather than flowing through it, as shown in Fig. 6.14.

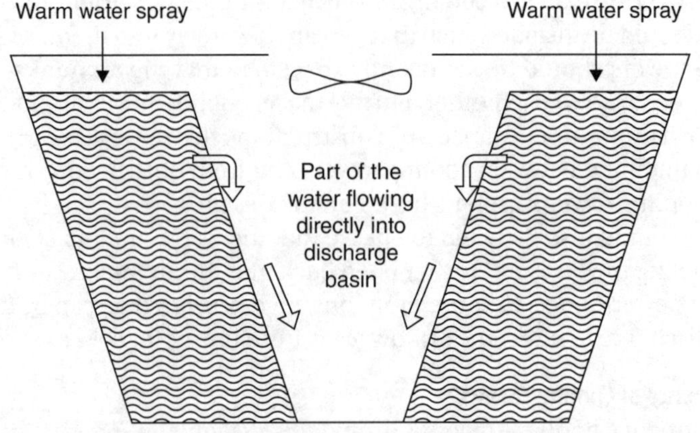

FIGURE 6.14 Water flow in defective cooling towers.

Since many cooling tower fans use belt drive systems, the tension of the belts should also be checked regularly to prevent belt slippage, which can result in a lower fan speed and a drop in cooling tower performance.

All the above aspects such as ineffective spray systems, damaged fill material, and lower fan speed can lead to poor cooling tower performance, which normally results in cooling towers not being able to provide water at the design temperature.

6.4   Low-Temperature Cooling Systems

Cooling systems with chilled water or other fluids, which use refrigeration systems to remove heat, are used for industrial cooling applications that require rapid cooling, low temperature cooling, or high heat removal rates.

A typical system is shown in Fig. 6.15 and consists of a refrigeration unit and liquid circulating pumps.

The refrigeration unit is typically a chiller or ice-making plant, which operates on the vapor compression or vapor absorption cycles. Chilled water is used as the working fluid for temperatures up to about 5°C, while other liquids such as glycol are used for subzero temperatures.

Refrigeration cycles: The vapor compression cycle has four mechanical components through which the refrigerant is circulated in a closed loop, as shown in Fig. 6.16.

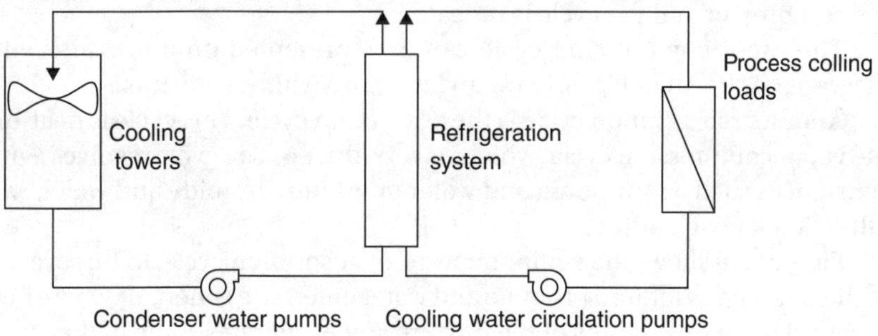

FIGURE 6.15   Typical arrangement of a liquid cooling system.

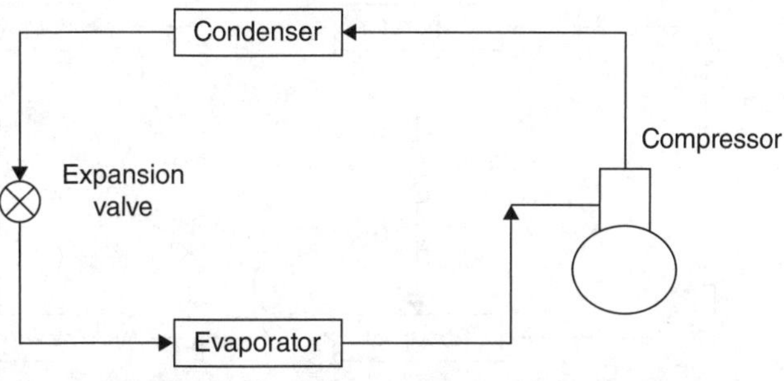

FIGURE 6.16   Vapor compression cycle.

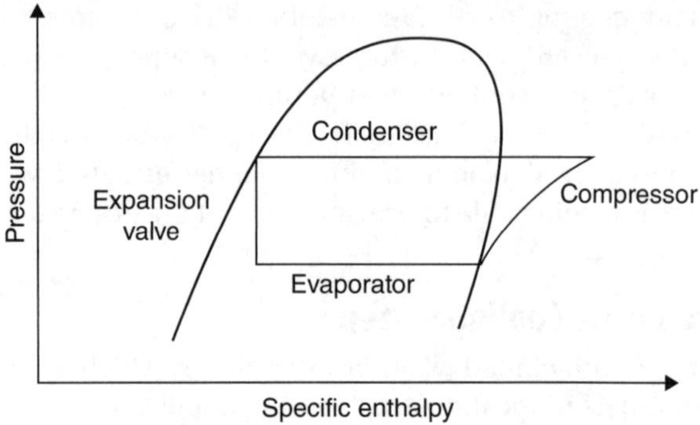

FIGURE 6.17   Pressure-enthalpy diagram for an ideal vapor compression cycle.

The refrigerant enters the compressor as low-pressure vapor and is compressed to high-pressure vapor. The high-pressure vapor then flows to the condenser, which is a heat exchanger where heat is rejected from the refrigerant and the refrigerant condenses from high-pressure vapor to high-pressure liquid. Next, the high-pressure liquid refrigerant flows through the expansion valve to the evaporator and becomes a low-pressure liquid refrigerant. In the evaporator, the liquid refrigerant evaporates at low temperature by absorbing heat from the surroundings. This is the cooling effect of the refrigeration system. The refrigerant then becomes low-pressure vapor and enters the compressor and the cycle is repeated.

The vapor compression cycle can be represented on a pressure–enthalpy (p-h) diagram, as shown in Fig. 6.17 for an ideal case without any losses.

Another refrigeration cycle is the absorption cycle. This cycle is heat-driven, unlike the vapor compression cycle, which is work-driven. The cycle involves a mixture of two substances such as ammonia and water or lithium bromide and water, which have an attraction for each other.

Figure 6.18 shows the components of an absorption cycle. In this cycle, the solution of the salt such as lithium bromide and water enters the generator where heat is applied. The heat breaks the bonds of the mixture and evaporates some of the water. The water vapor goes to the condenser where it is condensed from vapor to high-pressure liquid. This water, which acts as the refrigerant, passes through a metering valve and enters the

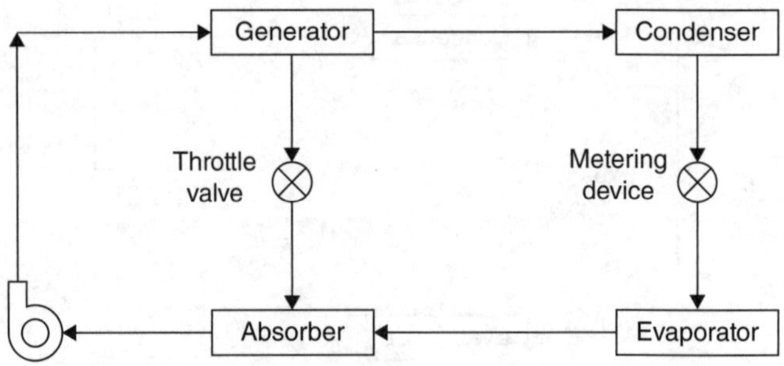

FIGURE 6.18   Components of the absorption cycle.

evaporator as low-pressure liquid. The water then evaporates in the evaporator by absorbing heat, providing the cooling effect of the cycle. The low-pressure vapor then enters the absorber where the strong salt solution returning from the generator absorbs it and is pumped as a weak solution back to the generator to complete the cycle.

The main difference between the absorption cycle and the vapor compression cycle is that the former uses a heat source for the generator as compared to doing "work" in the compressor of the vapor compression cycle. The absorption cycle is also less efficient than the vapor compression cycle. However, it is useful for applications where a free heat source (waste heat) is available. In such cases, the available heat can be used to produce steam or hot water for the absorption chiller.

Cooling capacity: The cooling capacity provided by a circulating fluid such as chilled water or glycol can be computed using the relationship

$$Q = m \times C_p \times \Delta T \tag{6.1}$$

where Q = cooling capacity in kW, m = mass flow rate of the fluid in kg/s, C_p = specific heat capacity of the fluid in kJ/kg · K, and ΔT = change in temperature in °C.

Cooling capacity can also be measured in refrigeration tons (RT) defined as the amount of cooling 1 ton (2000 lb) of ice can produce by melting over a 24 h period. Refrigeration of 1 RT has the ability to remove 288,000 Btu in a 24 h period.

Therefore, 1 RT = (2000 lb × 144 Btu/lb)/24 = 12,000 Btu/h

The conversion factor for the two different units is 1 RT = 3.517 kW.

Chiller-efficiency rating: Chiller efficiency is measured in terms of the amount of cooling produced by one unit of power or the amount of power used to produce one unit of cooling.

In the SI Metric system, chiller efficiency is measured in COP (coefficient of performance):

$$COP = \frac{kW \text{ refrigeration effect}}{kW \text{ input}}$$

In the English IP system, chiller efficiency is measured in kW/RT:

$$kW/RT = \frac{kW \text{ input}}{tons \text{ of refrigeration}}$$

6.4.1 Energy-Saving Measures for Low-Temperature Cooling Systems

The energy consumption in a chiller is mainly by the compressor motor and is dependent on the mass of refrigerant to be compressed, the compressor lift (difference between the evaporator and compressor pressures), and design characteristics of the cooling equipment.

Therefore, savings in chiller energy consumption can be achieved by reducing the compressor load (mass of refrigerant that needs to be compressed), reducing the compressor lift, and optimizing the operation based on compressor characteristics.

Reducing Compressor Load

Using Alternative Sources for Cooling   If a particular process does not require rapid cooling or a low-temperature cooling medium, water from a cooling tower or a natural heat sink such as a river can be used for heat removal, instead of chilled water or low-temperature liquid, which requires more energy intensive refrigeration systems.

Improving Insulation   Often the utility building where the chillers are installed is located a several hundred meters away from the equipment within the process plant that requires cooling. Therefore, the cooling liquid has to be transported over long distances and a significant heat gain can result due to poor or inadequate insulation of the distribution piping. Similarly, unnecessary heat gain can also take place if the vessels or heat exchangers are not adequately insulated. Care should be taken not only to insulate the main body of the equipment, but also the fittings such as valves and flanges as well.

Isolating Loads   As described earlier in the "Optimizing Pumping Systems" section, often a particular cooling system serves numerous heat exchangers serving various processes. Generally, these processes are operated in batches and require cooling only at specific times during each cycle. However, in most cases, the cooling fluid is allowed to flow through the heat exchanger even when heat removal is not necessary. This can lead to higher cooling load for the chillers due to excessive cooling and heat gain from piping and heat exchangers.

Therefore, isolation valves, ideally interlocked with the respective heat exchangers, should be used to isolate the flow of the cooling fluid when heat removal is not required by a particular equipment.

Reducing Compressor Lift

Using Water-Cooled Condensers   Chillers are classified as air cooled or water cooled according to the type of condenser used. Water-cooled chillers have shell and tube heat exchangers where the heat is rejected to the condenser cooling water. This warm condenser water is then pumped to cooling towers where the heat is rejected to the environment. In air-cooled chillers, a single fan or a number of fans are used to blow ambient air through the condenser and the heat is rejected to the air.

Water-cooled chillers operate at a lower condensing pressure than air-cooled chillers. The lower condensing pressure is due to the rejection of heat to condenser water, which is first cooled to a temperature a few degrees above the wet-bulb temperature of the ambient air (by the cooling towers). In air-cooled chillers, heat is directly rejected to the ambient air and the heat transfer is dependent on the dry-bulb temperature of the air. Furthermore, the heat transfer in the water-cooled shell and tube heat exchangers is better than in finned-type air-cooled condensers.

As shown in Fig. 6.19, the lower condensing pressure for water-cooled chillers leads to a lower compressor lift, which results in lower power consumption by the compressor. Therefore, water-cooled chillers are far more efficient than air-cooled chillers. The efficiency of water-cooled centrifugal chillers is about 0.5 to 0.6 kW/RT (COP of 6 to 7) compared to the typical efficiency of air-cooled chillers, which ranges from 1.0 to 1.2 kW/RT (COP of 2.9 to 3.5), operating at standard conditions of chilled water temperature of 6.7°C and condenser water supply temperature of 29.5°C.

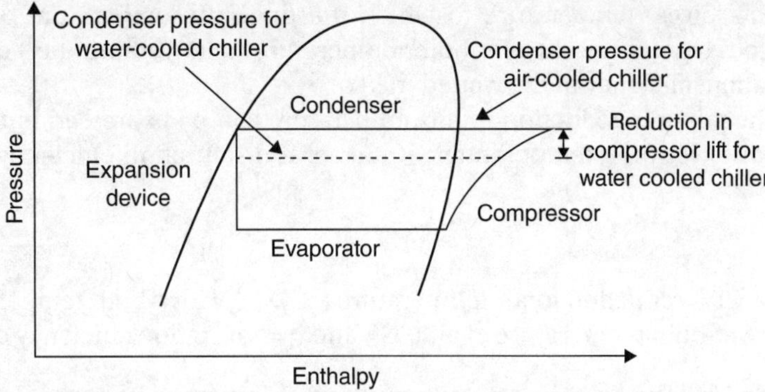

FIGURE 6.19   p-h diagram for water-cooled vs. air-cooled chillers.

Therefore, water-cooled chillers should be generally used instead of air-cooled chillers, unless the cost of water is relatively high or water is scarce. In applications where air-cooled chillers are used, they should be considered for replacement with water-cooled chillers. However, for practical reasons if it is not possible to replace an air-cooled chiller, a condenser precooler can be considered as an alternative (described below).

Condenser Pre-coolers   Condenser pre-coolers are evaporative coolers that use a wetted medium placed before the condenser to pre-cool the air stream before reaching the condenser coil. A typical arrangement is shown in Fig. 6.20. When warm ambient air passes through a wetted medium such as cellulose placed in a frame, some of the water evaporates from the medium, thereby reducing the temperature of the air stream.

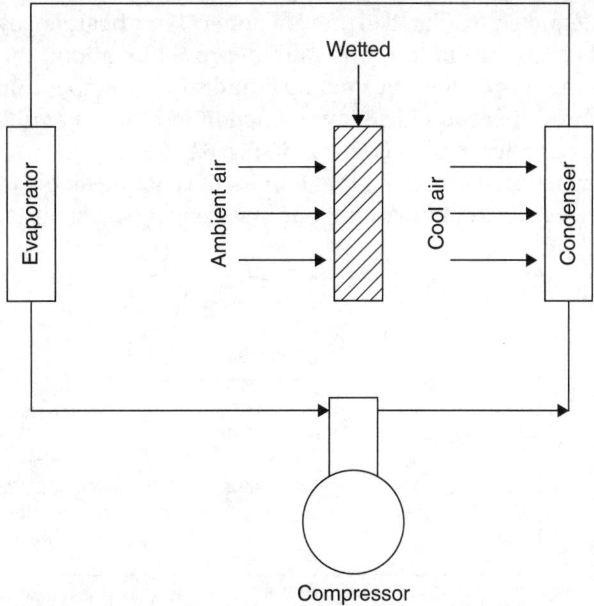

FIGURE 6.20   Arrangement of a condenser pre-cooler.

The lowest temperature to which the dry-bulb temperature of the air can be reduced depends on the wet-bulb temperature (relative humidity) of the air and the saturation efficiency of the wetted media.

The possible reduction in air temperature can be expressed in terms of the dry-bulb and wet-bulb temperatures of the air and saturation efficiency of the media as follows:

$$\Delta T = (T_{DB} - T_{WB}) \times \eta \tag{6.2}$$

where ΔT = reduction in air temperature (°C), T_{DB} = dry-bulb temperature of air (°C), T_{WB} = wet-bulb temperature of air (°C), and η = saturation efficiency of media (%).

Example 6.3   The average dry-bulb and wet-bulb temperatures of air are 35 and 28°C (60% RH), respectively. Estimate the possible reduction in the dry-bulb temperature of air, assuming the saturation efficiency of the media to be 80%.

Solution

$$\Delta T = (T_{DB} - T_{WB}) \times \eta$$
$$= (35 - 28) \times 80\%$$
$$= 5.6°C \quad \blacktriangle$$

Increasing Temperature of Cooling Fluid   The compressor lift, which is the pressure difference between the condenser and evaporator pressures, can be reduced by increasing the evaporator pressure as shown in Fig. 6.21.

The evaporator pressure can be maximized by operating the system to produce a cooling liquid at the highest temperature acceptable for the heat-removal process. To achieve this objective, it would also be necessary to ensure that the circulation piping system is adequately insulated to prevent unnecessary heat gain by the cooling medium.

The efficiency of chillers generally improves by about 2% to 4% for every 1°C increase in the set-point of the cooling liquid supply temperature.

The improvement in efficiency (reduction in kW/RT) at different operating temperatures for a typical chiller is shown in Fig. 6.22.

In manufacturing processes that require the cooling media to be at vastly different temperatures, instead of operating one system at a particular temperature to suit the

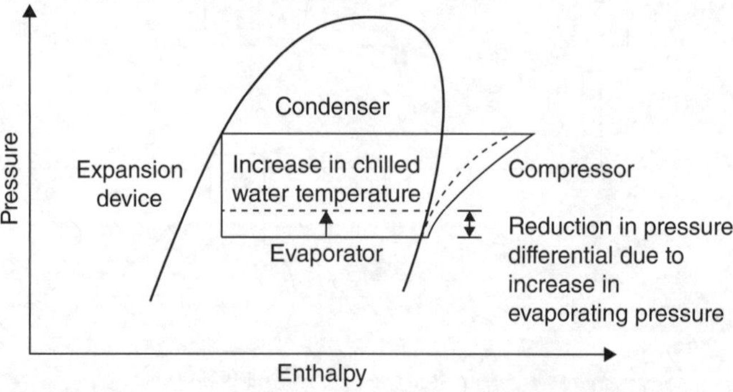

Figure 6.21   Reducing compressor lift by increasing the evaporator temperature.

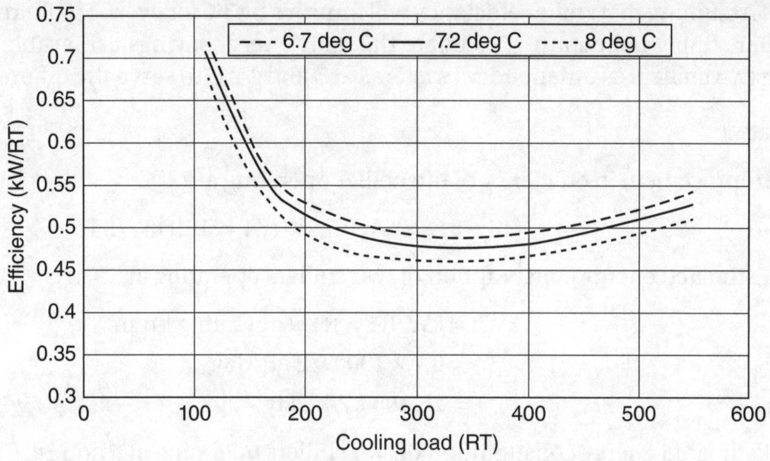

FIGURE 6.22   Chiller efficiency vs. operating temperature for a 550-RT-capacity chiller.

lowest value required, two or more systems can be operated at different temperatures, as illustrated in Fig. 6.23.

Example 6.4   A process cooling load of 1000 RT is served by two 500-RT-capacity chillers. The chillers provide chilled water at 5°C. However, only approximately half the cooling load requires chilled water at 5°C, while the remaining cooling load requires cooling water at only 14°C. The chiller plant operates 24 h a day and the operating efficiency of the chillers is 0.7 kW/RT.

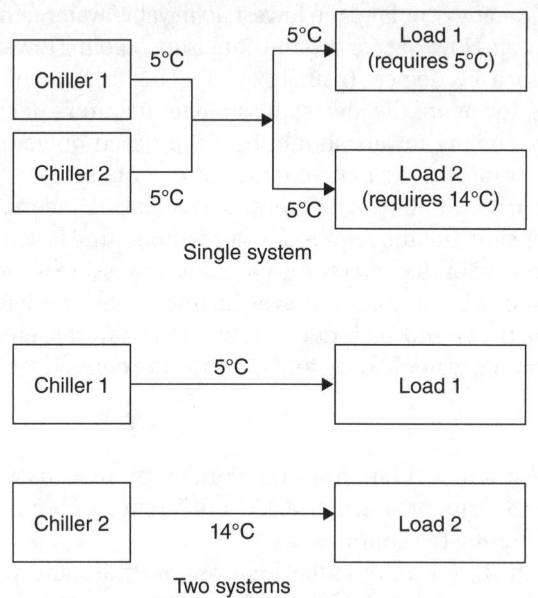

FIGURE 6.23   Single vs. multiple cooling systems.

Assuming that chiller efficiency will improve by 3% for every 1°C increase in chilled water supply temperature, estimate the daily energy savings achievable by operating the two chillers as independent systems at 5 and 14°C to serve the different loads.

Solution

Improvement in efficiency of the chiller operating at 14°C

$$= (14 - 5) \times 3\% \times 0.7 \text{ kW/RT} = 0.19 \text{ kW/RT}$$

Estimated energy consumption of two chillers operating at 5°C:

$$\text{kWh} = \text{kW/RT} \times \text{RT} \times \text{operating hours}$$
$$= 0.7 \times 1000 \times 24 \text{ h/day}$$
$$= 16{,}800 \text{ kWh/day}$$

Estimated energy consumption of two chillers operating at 5 and 14°C:

$$\text{kWh} = (0.7 \times 500 \times 24 \text{ h/day}) + [(0.7 - 0.19) \times 500 \times 24]$$
$$= 14{,}520 \text{ kWh/day}$$

$$\text{Energy savings} = (16{,}800 - 14{,}520) \text{ kWh} = 2280 \text{ kWh/day} \quad \blacktriangle$$

Reducing Condenser Pressure   Similar to increasing the evaporator pressure, the compressor lift can also be reduced by reducing the condenser pressure. The three main factors that affect the condenser pressure are the temperature of the condenser cooling fluid, flow rate of the cooling fluid, and the heat-transfer rate between the refrigerant and the cooling medium.

The most common condenser cooling medium is water, which is circulated through a cooling tower system to reject heat to the atmosphere. Since cooling towers remove heat mainly by evaporative cooling, the lowest achievable water temperature is the wet-bulb temperature of air. However, for economic reasons, cooling tower systems are designed to provide water at a few degrees (usually 2 to 3°C) higher than the wet-bulb temperature.

Therefore, to ensure the lowest possible temperature of condenser cooling water, the respective cooling towers should be sized based on actual operating conditions such as the flow rate of water and conditions of ambient air.

In addition, the cooling towers should be installed and maintained as explained in Sec. 6.3.1 to ensure that the required water temperature is achieved.

The second factor that affects the condenser pressure is the flow rate of the cooling water. Since the cooling water removes heat from the condenser by sensible cooling, the quantity of heat removed is dependent on the flow rate and the temperature difference of the cooling water leaving and entering the condenser.

$$Q = m \times C_p \times \Delta T \tag{6.3}$$

where Q = heat removal rate from the condenser, m = mass flow rate of water, C_p = specific heat capacity of water, and ΔT = difference in temperature of cooling water leaving and entering the condenser.

From Eq. (6.3), for a particular heat removal rate and water temperature at the condenser inlet, since the specific heat capacity is constant, when the mass flow rate reduces, the temperature of the water at the exit of the condenser increases. When the temperature of the cooling fluid at the discharge of the condenser increases, the condenser temperature and pressure increase, thereby leading to higher compressor lift.

For example, if a particular condenser is designed to operate with cooling water entering at 25°C and leaving the condenser at 30°C (ΔT of 5°C), and if the flow rate is reduced to half the design value, then ΔT will become 10°C and the temperature of cooling water leaving the condenser will increase to 35°C.

The third factor that affects the condenser pressure (and therefore the compressor lift) is the heat-transfer rate from the refrigerant in the condenser to the cooling fluid. The heat-transfer rate for a particular condenser is dependent on its design. However, the heat-transfer rate can reduce during operation due to scaling and fouling of the condensers due to increased resistance to heat flow. Therefore, condensers using water from cooling towers that are prone to scaling and fouling should be well maintained. Since it is not practical to manually clean condensers on a regular basis, automatic condenser tube cleaning systems should be used. The two main types of condenser cleaning systems are brush type and ball type.

Figure 6.24 shows a typical system using sponge balls for cleaning tubes. These sponge balls have a slightly larger diameter than the inside bore of the condenser tubes. The balls are circulated through the tubes at regular intervals. As the balls travel through the tubes, scale and fouling deposits are removed. After passing through the tubes, the balls are collected by a strainer. Thereafter, the balls are returned to the cleaning section for automatic cleaning and are injected back again after a set time.

Cooling Equipment Design Characteristics

The operating efficiency of cooling equipment also depends on compressor characteristics and controls, resulting in better efficiency when operating under particular load conditions.

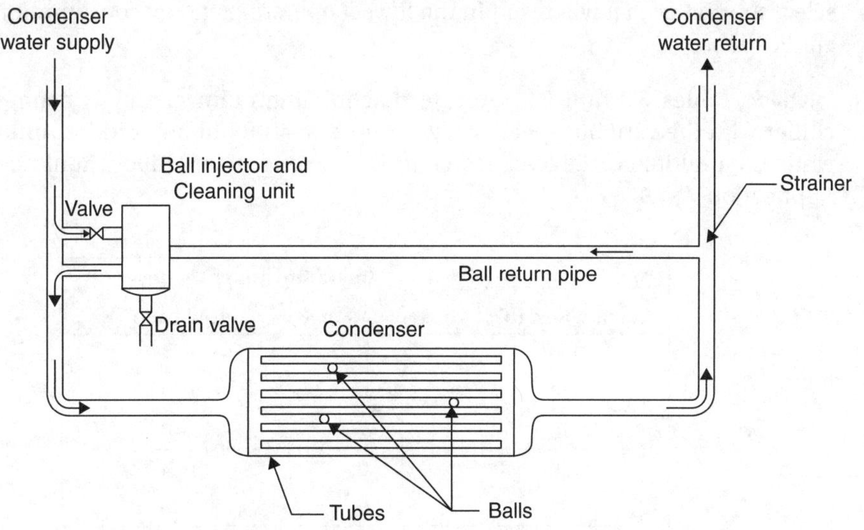

Figure 6.24   Arrangement of a typical ball-type condenser cleaning system.

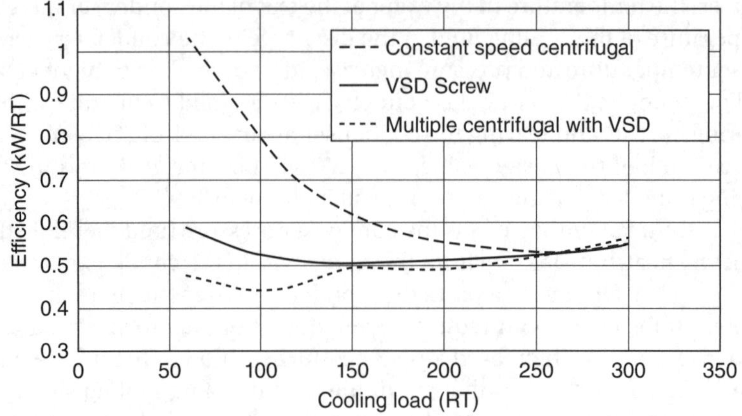

Figure 6.25 Part-load operating efficiency for three different 300-RT-capacity chillers.

Figure 6.25 shows how the operating efficiency of three different 300-RT-capacity chillers varies with loading. The chiller operating with a constant-speed centrifugal compressor has a good operating efficiency (low kW/RT) when the loading is above 80%. The chiller using a variable-speed compressor has a lower kW/RT below about 80% loading. Further, the operating efficiency of the chiller with multiple variable-speed centrifugal compressors is the best when loading is lower than about 50%.

Therefore, for optimum energy efficiency, the most suitable chiller for a particular operation should be selected based on the cooling load profile, as illustrated in Example 6.5.

Example 6.5 The cooling load profiles for two different cooling applications (X and Y) are shown in Table 6.1. Using the chiller part-load operating data provided in Fig. 6.25, select the chiller that will result in the lowest operating energy consumption for each application.

Solution Tables 6.2 and 6.3 indicate that for minimum energy consumption, the chiller with the variable-speed screw compressor should be used for Application X, while the multiple variable-speed centrifugal compressor chiller should be used for Application Y. ▲

	Operating hours per day	
Cooling load (RT)	Application X	Application Y
300	16	4
270	6	2
210	2	4
150	0	4
90	0	10

Table 6.1 Operating Hours for Example 6.5

Cooling load (RT)	Daily operating hours	Operating efficiency			Daily energy consumption		
		Constant-speed centrifugal compressor	Variable-speed screw compressor	Multivariable-speed centrifugal compressors	Constant-speed centrifugal compressor	Variable-speed screw compressor	Multivariable-speed centrifugal compressor
A	B	C	D	E	$A \times B \times C$	$A \times B \times D$	$A \times B \times E$
300	16	0.550	0.548	0.565	2640.00	2630.40	2712.00
270	6	0.537	0.535	0.531	869.94	866.70	860.22
210	2	0.548	0.513	0.496	230.16	215.46	208.32
150	0	0.62	0.505	0.498	0	0	0
90	0	0.856	0.536	0.448	0	0	0
				Total	3740.10	3712.56	3780.54

TABLE 6.2 Total Daily Energy Consumption for Cooling Applications X

Cooling load (RT)	Daily operating hours	Operating efficiency			Daily energy consumption		
		Constant-speed centrifugal compressor	Variable-speed screw compressor	Multivariable-speed centrifugal compressors	Constant-speed centrifugal compressor	Variable-speed screw compressor	Multivariable-speed centrifugal compressors
A	B	C	D	E	$A \times B \times C$	$A \times B \times D$	$A \times B \times E$
300	4	0.550	0.548	0.565	660.00	657.60	678.00
270	2	0.537	0.535	0.531	289.98	288.90	286.74
210	4	0.548	0.513	0.496	460.32	430.92	416.64
150	4	0.62	0.505	0.498	372.00	303.00	298.80
90	10	0.856	0.536	0.448	770.40	482.40	403.20
				Total	2552.70	2162.82	2083.38

TABLE 6.3 Total Daily Energy Consumption for Cooling Applications Y

6.5 Refrigeration Systems

Many industrial facilities such as food processing plants, dairy plants, breweries, and vegetable- and fruit-processing plants use refrigeration systems to cool and store products. Such systems include those used for rapid freezing of products and cold rooms used for the storage of products.

Common methods used for freezing products are as follows:

- Blast freezers where low temperature air is passed at high velocity over the products
- Contact freezing where the product is placed between freezer plates
- Spiral freezers where products move on a conveyor inside a cold chamber
- Belt freezers and freezer tunnels where cold air is forced at high velocity over a conveyor belt
- Immersion freezer where the product is immersed in a solution such as brine

Cold rooms are generally classified as chillers or freezers based on the operating temperature. Generally, chillers operate at a few degrees higher than the freezing point of water (2 to 5°C), while freezers operate at temperatures below the freezing point of water to as low as −35°C.

These refrigeration systems are similar to those described in Sec. 6.4 and operate on the vapor compression cycle. The arrangement of a typical system is shown in Fig. 6.26, which consists of one or more evaporators (usually installed in the refrigerated spaces such as cold rooms), compressors, and condensers. The refrigerants used for industrial refrigeration systems are ammonia (R717), R-22, and carbon dioxide (R744).

The refrigeration cycle can be single stage or multi stage. Figure 6.26 shows a single-stage system with liquid overfeed (explained later) used for multiple operating

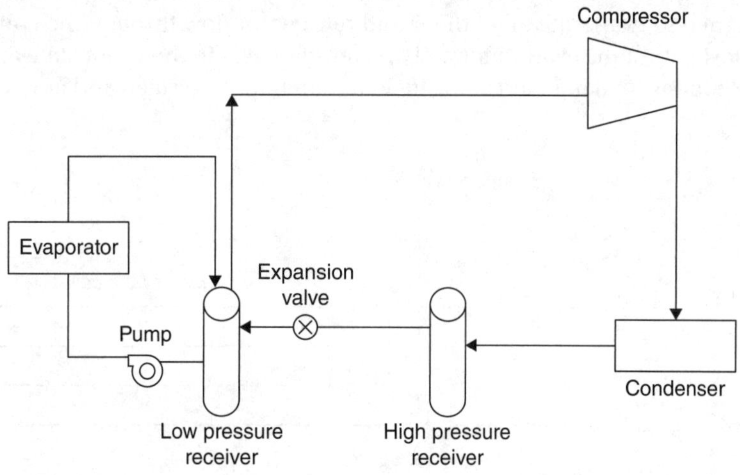

Figure 6.26   Single-stage system with liquid overfeed.

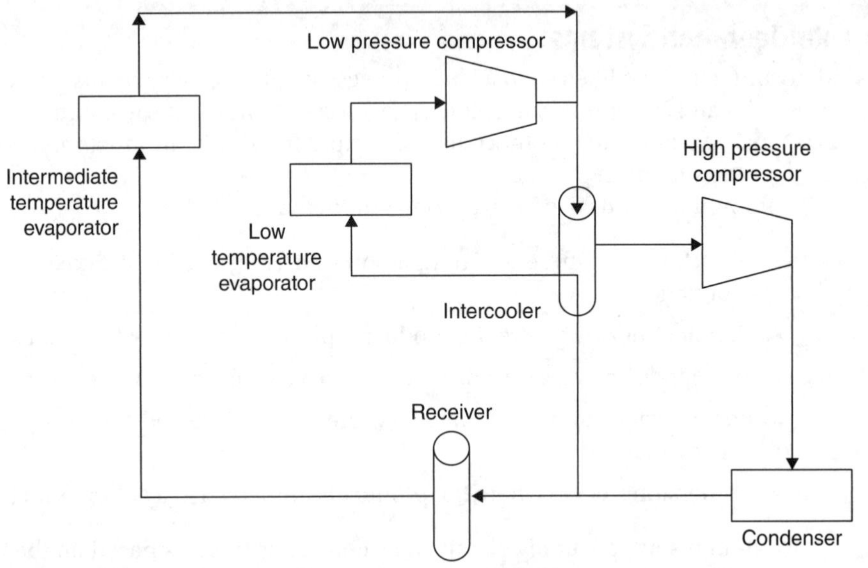

FIGURE 6.27   Multi-stage system with evaporators operating at two temperatures.

temperature applications, while Fig. 6.27 shows a multi-stage system used for low-temperature and intermediate-temperature applications.

The systems can also be classified as direct expansion (DX), flooded, or liquid overfeed, depending on the refrigerant flow in the system.

DX systems: In a DX system, liquid refrigerant (at high pressure) flows from the receiver to the evaporator coil through a thermal expansion valve. The thermal expansion valve controls the flow of refrigerant to the evaporator to maintain a set superheat value of the refrigerant leaving the evaporator coil, as shown in Fig. 6.28. DX systems are designed to evaporate all of the liquid refrigerant flow through the coil. Therefore, in the last part of the evaporator coil, the refrigerant is in the vapor state and cannot provide cooling, making them less efficient than liquid overfeed and flooded coils.

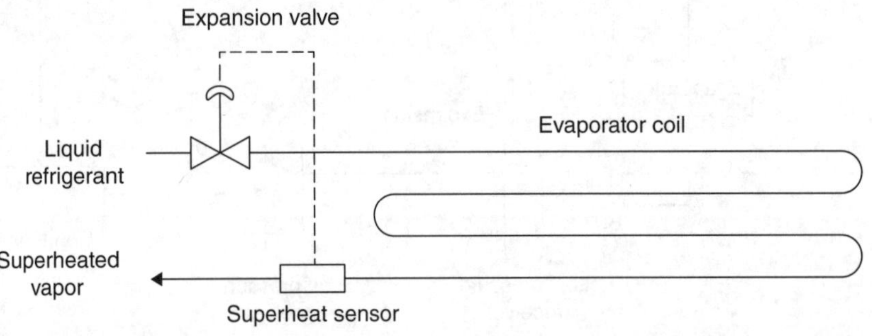

FIGURE 6.28   Arrangement of a DX system.

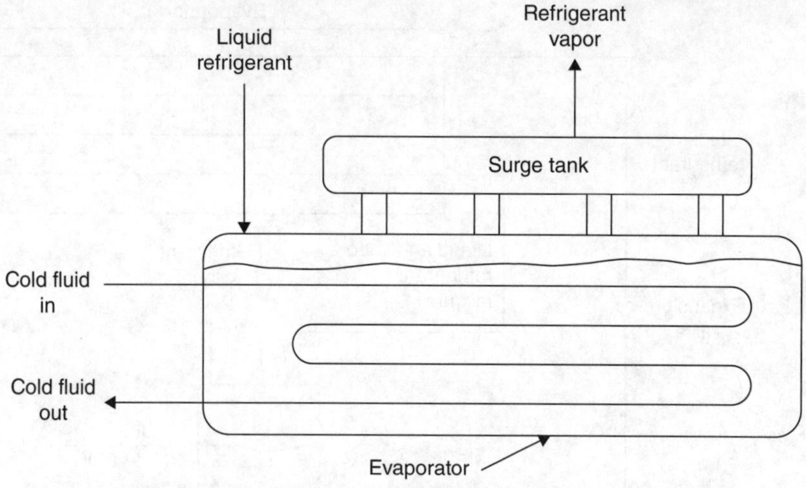

FIGURE 6.29   Arrangement of a flooded-type evaporator.

Flooded systems: In flooded-type evaporators, liquid refrigerant is stored in an accumulator vessel located above the evaporator coil. The refrigerant flows by gravity to the coil. The refrigerant vapor bubbles formed during the evaporation process rise through the coil into the accumulator. Thereafter, the vapor is drawn into the compressors. A pressure regulator is installed on the pipe between the accumulator and the compressor to control the coil pressure and rate of evaporation (Fig. 6.29).

Liquid-overfeed systems: In liquid-overfeed or recirculation systems, liquid refrigerant is pumped from a low-pressure receiver to the evaporator coils. The flow rate is set to provide the evaporator coils with more liquid than is required to evaporate and satisfy the load, resulting in "overfeed" or recirculation of the refrigerant (Fig. 6.30). The refrigerant returning from the coil to the receiver is a mixture of liquid and vapor. The vapor rises to the top of the receiver and is extracted by the compressor. The capacity of each evaporator coil is controlled by a solenoid valve that can be opened or closed.

Although these refrigeration systems are similar to process cooling systems operating on the vapor compression cycle, they are usually more energy intensive due to the relatively low evaporator operating temperatures.

6.5.1   Energy-Saving Measures for Refrigeration Systems

The energy-saving measures are similar to those described in Sec. 6.5 and involve reducing the refrigeration load, increasing the evaporator pressure, reducing the condenser pressure, and optimizing compressor operation.

Reducing Refrigeration Load

In industrial refrigeration systems, the total load is dependent on the starting and end temperatures of the product being cooled and the heat gain thereafter during storage. Hence, if the temperature of the product is high when reaching the refrigeration

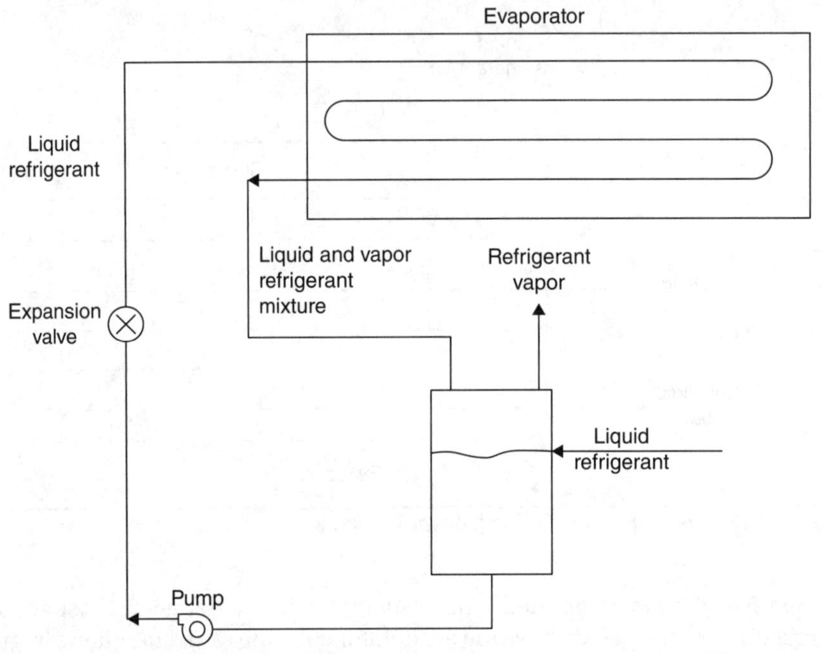

FIGURE 6.30   Arrangement of a liquid-overfeed system.

system, alternative cooling systems can be explored to pre-cool the product before being refrigerated. Such alternative cooling systems can be ambient cooling or using other less energy intensive cooling media like cooling tower water or chilled water.

During the storage of products in a cold room, the heat gain is dependent on factors such as the storage temperature, insulation of the storage space, air leakages between the cold space and external environment, and heat added by lighting installed inside the storage space.

The lower the storage temperature, the higher will be the temperature gradient between the outside and inside of the cold space, which is the potential for heat gain. Therefore, the cold space temperature should be set to the highest possible value acceptable for the application. This will not only reduce the refrigeration load, but also help to improve the operating efficiency of the compressors (explained later). Some typical storage temperatures are summarized in Table 6.4.

Once the product being stored reaches the storage temperature, the refrigeration system operates continuously only to remove heat gain from external and internal sources of heat. The main internal source of heat is lighting. Therefore, energy-efficient lighting such as high-pressure sodium, metal halide, and fluorescent lamps, which emit less heat, should be used, while the usage of lighting should be minimized by switching them off when not required. Occupancy sensors can be used to switch off most of the lighting, leaving a few to be controlled manually.

Due to the high temperature gradient between the refrigerated space and the ambient air, good insulation should be used to minimize heat gain through the exterior

Product	Typical storage temperature (°C)
Cheese	0 to 1
Poultry	−2 to 0
Pork and beef	0 to 1
Fish	−0.5 to 2
Butter	−23
Milk (raw)	0 to 4
Milk (pasteurized)	4 to 6
Vegetables (beet, carrots, cabbage, mushroom, etc.)	0
Fruits (berries, apricot, grapes, etc.)	−0.5 to 0

TABLE 6.4 Typical Product Storage Temperatures

surfaces of the storage space and refrigerant pipes. In addition, air leakage between the cold space and the exterior should be minimized by sealing door seals and only opening the doors when required. In applications where the doors need to be frequently opened, automatic doors or strip curtains should be installed.

Reducing Evaporator Pressure

The operating efficiency of a compressor in refrigeration systems improves when the suction pressure (and temperature) is increased. Typically, an increase of 1°C in suction temperature improves the compressor operating efficiency by about 3% to 4%. Therefore, refrigeration systems should be operated at the maximum possible operating temperature.

In applications where a refrigerated space is used for storing different products at different times, the operating temperature should, ideally, be changed every time the product being stored is changed. For instance, if a cold room normally used to store poultry at −2°C is now used to store beef, the temperature can be set at 1°C which will improve the compressor efficiency by about 10%.

In systems serving multiple applications with a wide range of operating temperatures, separate systems to serve low-temperature and high-temperature applications can improve compressor efficiency, as explained in the "Reducing Condenser Pressure" section.

Optimizing Evaporator Fans

Evaporators consist of coils where the liquid refrigerant evaporates absorbing latent heat of vaporization and becomes refrigerant vapor, thereby providing cooling. Evaporators have fans, in addition to the coil, to circulate air between the coil and the cooled space, as shown in Fig. 6.31.

Although the rated capacity of evaporator fans is relatively low compared to the compressor power, since they operate 24 h a day, they can account for a significant amount of energy consumption. The first step is to select an evaporator with a relatively

FIGURE 6.31 Typical evaporator (courtesy of Evapco).

lower pressure drop to minimize energy consumption by the fan. The fan power in kW divided by the cooling capacity in RT should not exceed 0.2 kW/RT. To reduce energy consumption during operation, VFDs can be installed together with a suitable temperature sensor and controller to reduce the fan speed or completely switch them off when the desired space temperature is reached.

Defrosting

Most evaporator coils operate at below freezing point and frost will form over the surface. If the frost is not removed, it will reduce the cooling provided by the coil and prevent air flowing through it. Therefore, periodically, the ice formed on the evaporator coil is removed by a defrosting cycle. The most common defrosting method in industrial systems is hot gas defrost where hot refrigerant vapor from the compressor is passed through the evaporator coil to melt the ice.

The frequency of the defrost cycle is usually based on a time cycle. Although automatic controls are available for activating defrosting using frost thickness, pressure difference across the coil, and temperature drop across the coil, due to practical difficulties faced in operating such systems, the most common method is to use a timer cycle.

Since a significant amount of heat is added to the refrigerated space during defrosting, the activation and duration of the defrosting cycle should be minimized. In applications where the load does not vary significantly, the activation time and duration for defrosting can be set manually by observing the time taken for ice to build up on the evaporator coil.

Reducing Condenser Pressure

Condensers used for refrigeration systems can be air cooled, water cooled with cooling towers, and evaporative. Air-cooled condensers consist of refrigerant coils with circulation fans, as shown in Fig. 6.32. The condensing pressure and therefore the temperature is dependent on the dry-bulb temperature of air.

Figure 6.32 Arrangement of an air-cooled condenser.

Water-cooled condensers consist of shell and tube condensers or plate heat exchangers where the refrigerant is condensed into a liquid using water from cooling towers. As explained earlier in this chapter, the temperature of cooling tower water used for rejecting heat from the condenser can reach close to the wet-bulb temperature of ambient air.

Evaporative condensers are commonly used in industrial refrigeration systems. They are essentially a combination of a cooling tower and refrigerant condenser, as shown in Figs. 6.33 and 6.34. In evaporative condensers, the condenser is placed inside

Figure 6.33 Image of an evaporative condenser (courtesy of Evapco).

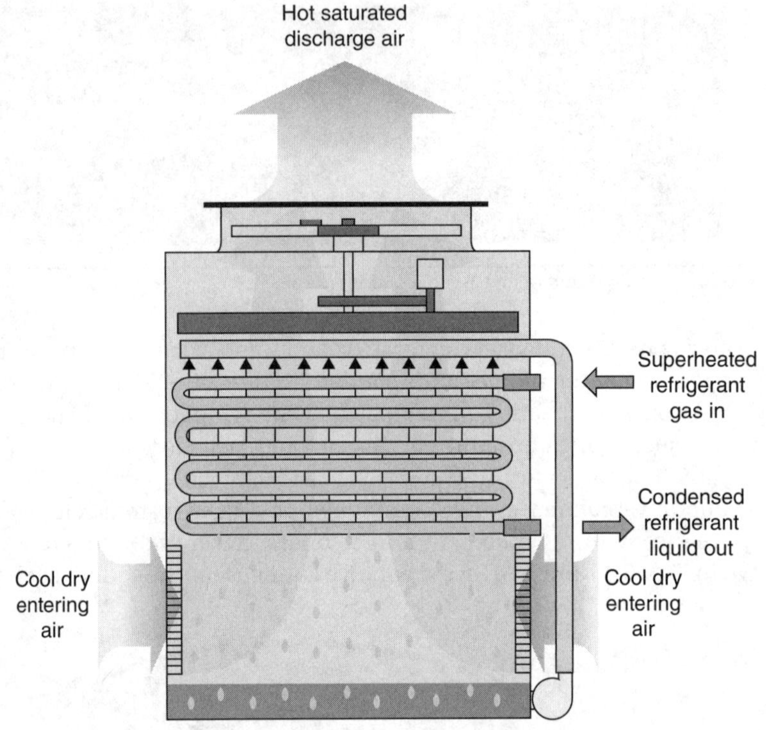

Figure 6.34   Operation of an evaporative condenser (courtesy of Evapco).

a tower where water is circulated over it to provide heat rejection. Water is supplied to make up for the water evaporated to provide heat rejection.

The operating efficiency of compressors is dependent on the compressor lift, which is the difference between the suction and discharge pressures. The lower the compressor lift, the better the operating efficiency of compressors, as explained in the "Reducing Condenser Pressure" section. The compressor lift can be reduced by increasing the suction pressure and reducing the condensing pressure. The refrigerant condensing pressure is dependent on the temperature of the cooling medium. Hence, water-cooled condensers and evaporative condensers, which reject heat at temperatures close to the wet-bulb temperature of air, result in lower condensing pressures compared to air-cooled condensers.

The efficiency gain when using water-cooled systems instead of air-cooled condensers is dependent on the dry-bulb and wet-bulb temperatures of ambient air. In general, improvement in compressor efficiency of 10% to 20% can be achieved by using water-cooled systems.

As explained in the "Optimizing Cooling Towers" section, the capacity of cooling towers can be varied (using VFDs) during part-load operation and periods of favorable outdoor conditions to minimize energy consumption. Similarly, VFDs can be used to vary the capacity of fans used for air-cooled condensers and evaporative condensers by installing VFDs with suitable temperature sensors and controllers to vary the condenser pressure in relation to the outdoor temperature (floating head control).

Minimizing Compressor Lift

The operating efficiency of compressors is dependent on the compressor lift, which is the difference between the suction and discharge pressures. The compressor lift can be reduced by increasing the suction pressure and reducing the condensing pressure.

The compressor suction pressure is dependent on the evaporator pressure (operating temperature of the refrigerated space). Therefore, the refrigerated space temperature should be set at the maximum value required for the application. As explained in Sec. 6.4.1, a 1°C increase in temperature will improve the operating compressor efficiency by 3% to 4%.

In some installations, one refrigeration system serves multiple applications that require different operating temperatures. Since the system is common to all the users, the compressor suction pressure has to be set to meet the lowest required temperature among all the users. The suction pressure of the evaporators operating at a higher temperature is maintained by throttling the refrigerant flow by artificially inducing a pressure loss, as shown in Fig. 6.35. Therefore, for example, if an installation serves a freezer system operating at −20°C and an ice cream production system that requires −35°C, using separate systems would result in better efficiency for the one serving the freezers operating at 10°C higher. Based on the expected improvement in efficiency of about 3% for 1°C rise in evaporator temperature, the improvement in efficiency in this case will be about 30% and can have a significant impact on the overall energy consumption if the freezer load is relatively high.

Similarly, the compressor lift can be reduced by reducing the condensing pressure, as explained in the "Reducing Compressor Lift" section. Reduction in condensing pressure can be achieved by using water-cooled systems instead of air-cooled systems, increasing the surface area available for heat rejection (larger condensers), and increasing the airflow rate in cooling towers. However, using water-cooled condensers instead of air-cooled ones and increasing the area available for heat transfer will result in a higher capital cost.

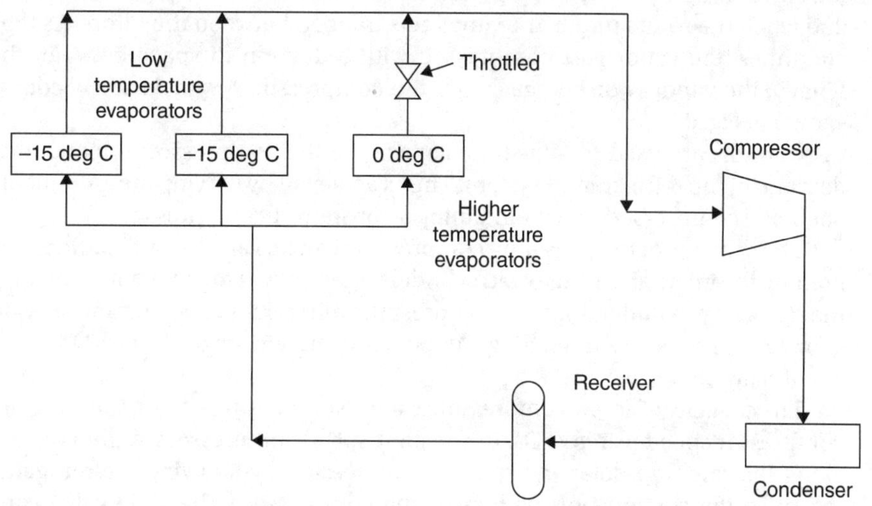

FIGURE 6.35 Arrangement of a single system serving multiple applications.

FIGURE 6.36   Cut-away image of a reciprocating compressor (courtesy of Bitzer).

Other means of reducing compressor lift involve reducing the pressure losses in the refrigerant piping system, which can be achieved by reducing the refrigerant pipe length and increasing pipe sizing.

Optimizing Compressor Operation

The most common types of compressors used for refrigeration applications are reciprocating and screw. Reciprocating compressors consist of multiple pistons and cylinders. Figure 6.36 shows a typical reciprocating compressor where the refrigerant vapor is compressed by the pistons attached to a rotating crankshaft.

Screw compressors are mainly twin screws that have a male and female rotor rotating and meshing together. Refrigerant vapor is filled into the space between the rotors, and when the rotors rotate, the vapor gets trapped between the rotors. As the rotation continues, the vapor gets compressed with reduction in space between the rotors. Finally, the vapor is discharged from the compressor. A typical screw compressor is shown in Fig. 6.37.

Controls are used to adjust the capacity of the compressor to match the varying demand of the refrigeration system. This is achieved by varying the amount of refrigerant being compressed and maintaining a constant suction pressure.

The capacity of reciprocating compressors can be varied by unloading cylinders by holding the inlet valve of selected cylinders open so that the movement of the piston in the respective cylinder does not compress the refrigerant vapor. In compressors with 4, 8, or 12 cylinders, the unloading can be performed in stages from 100% to 75%, 50%, and, finally, 25%.

The capacity of screw compressors can be varied using sliding valves or varying the compressor speed using VFDs. In constant-speed compressors, a slide valve is moved along the rotors to delay the start of compression by allowing the refrigerant vapor return to the suction port without being compressed. The slide valve can provide capacity control down to about 10% of the rated capacity.

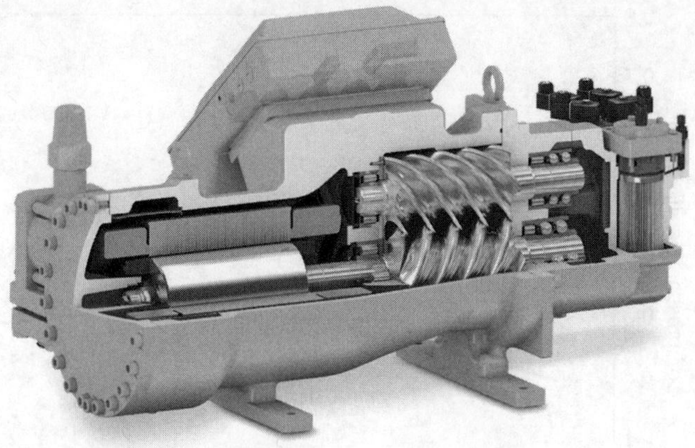

FIGURE 6.37 Cut-away image of a screw compressor (courtesy of Bitzer).

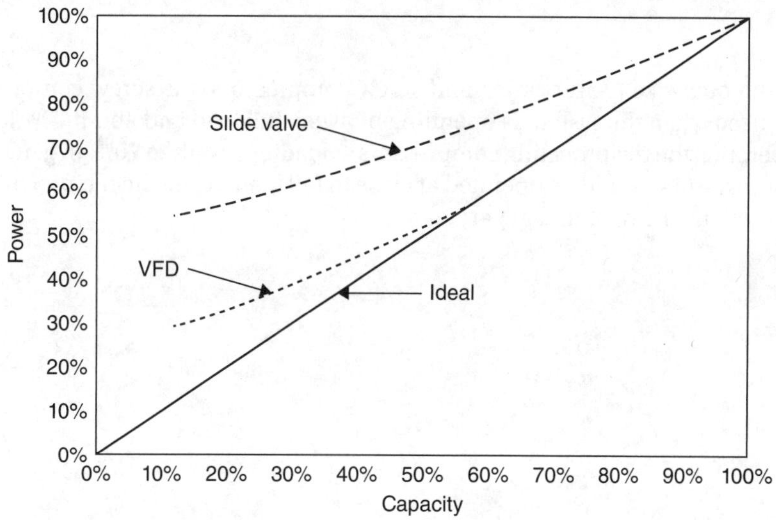

FIGURE 6.38 Performance comparison for slide valve vs. speed control at part load.

The other alternative is to use a VFD to reduce the speed of the compressor. When the operating speed is reduced, the volume of the refrigerant vapor compressed is also reduced. Reducing compressor speed instead of using slide valves to control the capacity of screw compressors is much more energy efficient, as shown in Fig. 6.38.

A comparison of performance between reciprocating and screw compressors (fixed speed) at a varying load is shown in Fig. 6.39. As shown in the figure, screw compressors have superior efficiency at full load, while reciprocating compressors have much better efficiency than screw compressors at part load. Therefore, in installations

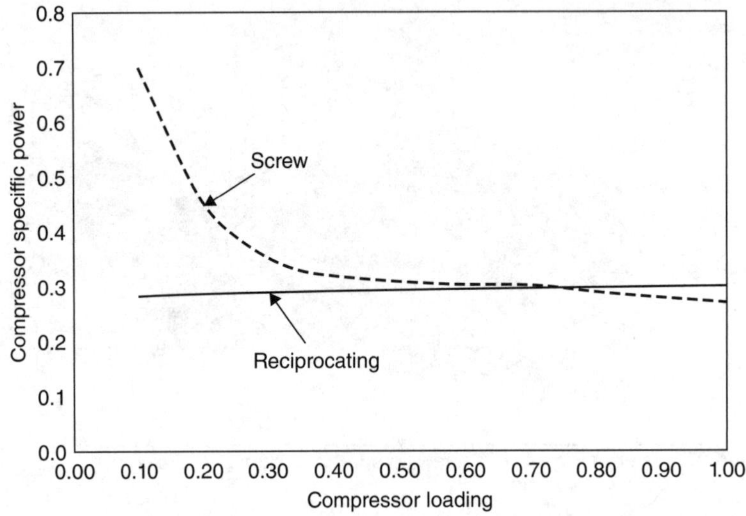

FIGURE 6.39   Comparison between reciprocating and screw compressors.

with equal-size reciprocating and screw compressors, the screw compressors should be used when the system is operating between full load and about 70% loading, while operating the reciprocating compressors at loading less than 70%. In general, the screw compressors should be operated at close to full load, while the reciprocating compressors should be operated at part load.

CHAPTER 7

Compressed Air Systems

7.1   Introduction

Compressed air is widely used in industrial applications such as pneumatic tools, control systems, operation of machinery, and manufacturing processes. Compressed air is essential for the operation of industrial plants and is often considered the "fourth utility" after electricity, water, and steam.

Compressed air systems are highly energy intensive and account for a significant portion of energy consumed in many plants. Most of the energy used in compressed air systems is wasted as heat and only about 10% of the input energy is delivered in usable form as compressed air. The typical energy flow for a compressed air system is shown in Fig. 7.1.

Normally, the cost of operating a compressed air plant is more significant than the first cost of the actual air compressors. Therefore, it is essential to optimize the design of compressed air plants so that they can operate efficiently to minimize operating costs.

7.2   Typical System Components

Compressed air systems comprise air compressors, filters, dryers, storage receivers, distribution systems, and end users (Fig. 7.2). The various components can be broadly divided into supply-side and demand-side components where the supply side consists of air compressors, filters, dryers, and receivers, while the demand side comprises the distribution system, storage, and the end users.

7.3   Free Air Delivery

Free air delivery (FAD) is a measure of air taken into the compressor at atmospheric temperature and pressure. The capacity of compressors and most end-user equipment is normally rated in FAD, which is a common reference point.

If a compressor is rated 5 m³/min free air at 9 bar, it means that 5 m³/min of air is taken at the compressor inlet to produce compressed air at 9 bar pressure.

FAD can be computed from the volume flow rate at the discharge of the compressor (Fig. 7.3) and using the ideal gas law as described below.

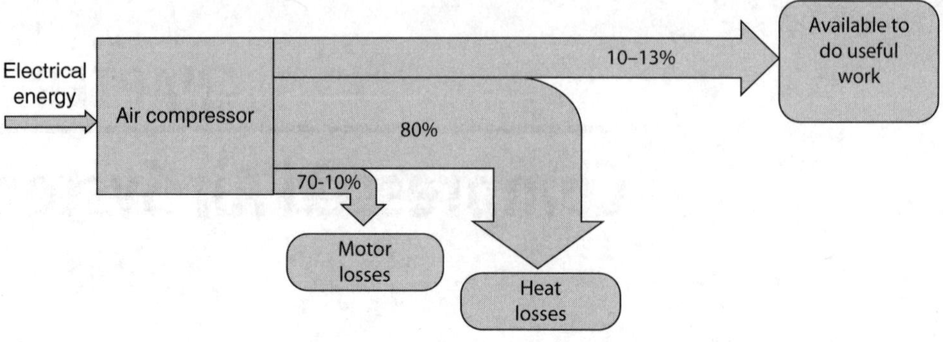

FIGURE 7.1   Energy flow for a compressed air system.

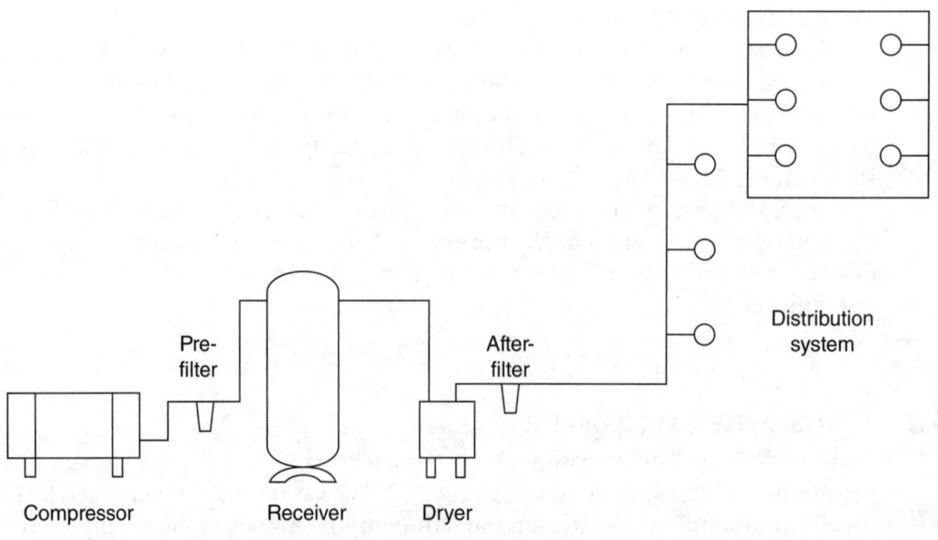

FIGURE 7.2   Arrangement of a typical compressed air system.

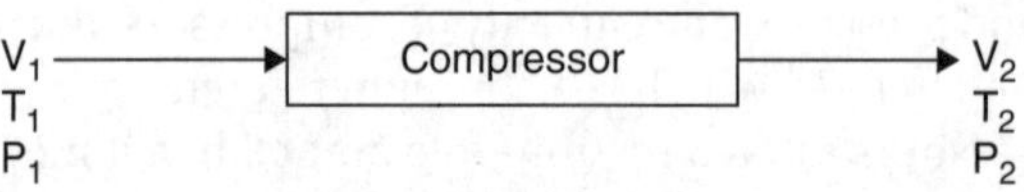

FIGURE 7.3   Diagram for computing FAD for a compressor.

For an ideal gas:

$$(P_1 \times V_1)/T_1 = (P_2 \times V_2)/T_2 \tag{7.1}$$

where P_1 and P_2, V_1 and V_2, and T_1 and T_2 are the absolute pressures, volumes, and absolute temperatures of air at the inlet and outlet of the compressor, respectively.

Therefore, for FAD:

$$V_1 = (V_2 \times P_2 \times T_1)/(T_2 \times P_1) \tag{7.2}$$

Example 7.1   The volume flow rate at the discharge of the compressor is measured to be 0.05 m³/s. The absolute discharge pressure and temperature are 6 bar and 35°C, respectively. Compute the corresponding FAD for the compressor, taking the ambient air temperature and absolute pressure to be 30°C and 1 bar, respectively.

Solution   Free air delivery:

$$V_1 = (V_2 \times P_2 \times T_1)/(T_2 \times P_1)$$
$$= [0.05 \times 6 \times (273.15 + 30)]/[(273.15 + 35) \times 1]$$
$$= 0.295 \text{ m}^3/\text{s} \quad \blacktriangle$$

7.4   Standard Conditions

Since the volume of air is dependent on the pressure, temperature, and relative humidity (measure of moisture in the air – discussed later), the parameter SCFM (standard cubic feet per minute), which refers to the volume flow rate at a set of standard conditions, is used to quantify the airflow rate.

Some of the commonly used "standard" conditions are summarized in Table 7.1.

The airflow rate at the actual conditions can be converted into that at standard conditions and vice versa using the ideal gas law.

From Eq. (7.1):

$$(\text{ACFM} \times P_a)/T_a = (\text{SCFM} \times P_s)/T_s \tag{7.3}$$

where ACFM = actual airflow rate in ft³/min, P_a = absolute pressure at actual conditions, T_a = absolute temperature at actual conditions, SCFM = airflow rate at standard conditions in ft³/min, P_s = absolute pressure at standard conditions, and T_s = absolute temperature at standard conditions.

Standard/usage	Temperature (°C)	Absolute pressure (bar)	Relative humidity (%)
CAGI & ASME	20	1.01	36
US	15.6	1.01	0
Europe & ISO	20	1.01	0

TABLE 7.1   Common Standard Conditions Used for Compressed Air

Compressed air also contains moisture and the relationship (7.3) can be modified, as given below, to include the vapor pressure exerted by the moisture in the air, to account for moisture in the air.

$$[\text{ACFM} \times (P_a - VP_a)]/T_a = [\text{SCFM} \times (P_s - VP_s)]/T_s \tag{7.4}$$

where VP_a and VP_s are the vapor pressures at the actual and standard conditions, respectively.

Relative humidity, RH, is the ratio of vapor pressure (VP) to saturated vapor pressure (SVP):

$$VP = RH \times SVP$$

where SVP = saturated vapor pressure.

Therefore, Eq. (7.4) can be expressed using the RH at actual and standard conditions, RH_a and RH_s, as

$$(\text{ACFM} \times [P_a - (RH_a \times SVP_a)])/T_a = (\text{SCFM} \times [P_s - (RH_s \times SVP_s)])/T_s \tag{7.5}$$

Hence, if the airflow rate is known at one set of conditions, the airflow rate at a different set of conditions can be computed using the above relationship (7.5).

7.5   Utilization Factor

The utilization factor (also called the load factor) for a compressed air user is the ratio of the actual compressed air consumption to the maximum continuous consumption.

$$\text{Utilization factor} \quad \frac{\text{actual air consumption in 24 h}}{\text{maximum continuous air consumption in 24 h}} \tag{7.6}$$

Example 7.2   The rated compressed air usage by a machine is 0.01 m³/s. If the actual compressed air usage by this machine is measured to be 260 m³/day, calculate the utilization factor.

Solution

$$\text{Utilization factor} = 260/(0.01 \times 3600 \times 24) = 0.3 \quad \blacktriangle$$

Utilization factor is used to estimate the total compressed air requirements when designing a new plant. Although many industrial plants have a large number of pneumatic tools and users, most of them may not operate simultaneously. Therefore, to prevent oversizing of the system, utilization factors for the various users can be used to compute the approximate average total compressed air consumption, as illustrated in Example 7.3.

Similarly, the utilization factor for a compressor can be expressed as the ratio of the actual compressed air output in a period of time to the output when operating at maximum capacity for the same period of time.

Equipment type	A	B	C
Air demand (FAD)	5 cfm	3 cfm	12 cfm
Working pressure	40 psig	60 psig	90 psig
Quantity	2	1	2
Utilization factor	50%	30%	20%

TABLE 7.2   Data for Example 7.3

Example 7.3   Compressed air requirements for a new system are summarized in Table 7.2. Estimate the FAD for a suitable compressor for this application.

Solution   The effective air demand for each type of equipment can be computed as follows:

$$\text{Equipment A} = 5 \times 2 \times 0.5 = 5 \text{ cfm}$$
$$\text{Equipment B} = 3 \times 1 \times 0.3 = 0.9 \text{ cfm}$$
$$\text{Equipment C} = 12 \times 2 \times 0.2 = 4.8 \text{ cfm}$$
$$\text{Total estimated demand} = 5 + 0.9 + 4.8 = 10.7 \text{ cfm} \quad \blacktriangle$$

7.6   Types of Compressors

The two main types of compressors are positive displacement and dynamic (Fig. 7.4). In positive-displacement compressors, a fixed volume of gas is compressed in a compression chamber where the volume of the gas is mechanically reduced. In such compressors, the gas flow is constant at a particular speed, irrespective of the discharge pressure. Typical positive-displacement compressors are reciprocating, screw, scroll, and rotary sliding vane, based on their respective operating characteristics.

In dynamic compressors, kinetic energy is imparted to a continuous flow of gas by a single rotor or multiple rotors operating at high speed. The imparted kinetic energy is later converted into potential energy in the discharge volute or diffuser of the compressor. Centrifugal and axial-flow compressors are dynamic compressors.

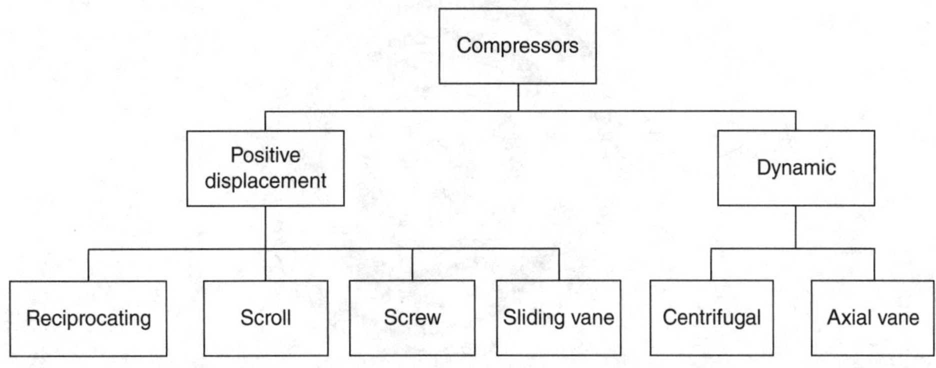

FIGURE 7.4   Different types of compressors.

Figure 7.5 Image of a reciprocating compressor.

7.6.1 Reciprocating Compressors

Reciprocating compressors use pistons and connecting rods driven by a crankshaft (Fig. 7.5). The crankshaft is driven by a motor. Compression is achieved by reducing the volume within the cylinder due to the movement of the piston. Normally, reciprocating compressors have a number of cylinders in one unit. For high-capacity applications, multistage units with multiple compressors are used. In such compressors, the gas is compressed to an intermediate pressure in the first stage, and then cooled to a lower temperature before being compressed further in the second stage.

7.6.2 Scroll Compressors

Scroll compressors use two inter-fitting, spiral-shaped scroll members as shown in Fig. 7.6. One scroll rotates while the other remains stationary. Due to the profile of the

Figure 7.6 Arrangement of a scroll compressor.

scrolls, the gas drawn in through the inlet port is compressed between the scrolls during rotation and then discharged at the discharge port.

7.6.3   Screw Compressors

Screw compressors consist of a set of male and female helically grooved rotors and compression is achieved by direct volume reduction due to the rotation of the rotors. The gas is taken in at the inlet port and then compressed during the rotation of the rotors and finally discharged at the discharge port. The image of a typical screw compressor is shown in Fig. 7.7.

7.6.4   Sliding-Vane Compressors

Sliding-vane compressors consist of a rotor with slots and sliding vanes placed in them as shown in Fig. 7.8. The rotor is eccentrically arranged within the housing, providing a crescent-shaped swept area between the intake and discharge ports. The rotor vanes are forced out from the slots up to the housing wall by centrifugal force during rotation.

FIGURE 7.7   Image of a screw compressor.

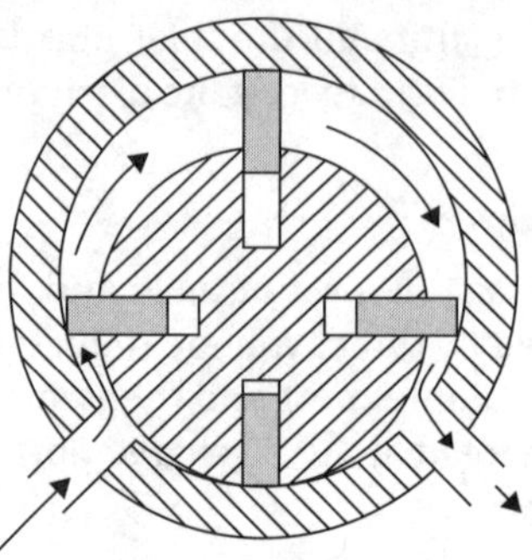

FIGURE 7.8   Cut-away of a rotary sliding-vane compressor.

Gas enters through the intake port, and gets trapped between two sliding vanes, the rotor, and the housing. As the rotor turns, the volume occupied by the trapped gas reduces to a minimum where it is exhausted at the discharge port.

7.6.5   Centrifugal Compressors

Centrifugal compressors consist of a single impeller or a number of impellers mounted on a shaft and rotating at high speed inside a housing. The gas to be compressed enters the impeller in the axial direction and is discharged radially at high velocity. The velocity pressure is then converted into static pressure in the diffuser. Since the part-load efficiency of centrifugal compressors is low, they are normally used for base-load applications. The image of a centrifugal compressor is shown in Fig. 7.9.

FIGURE 7.9   Image of a centrifugal compressor.

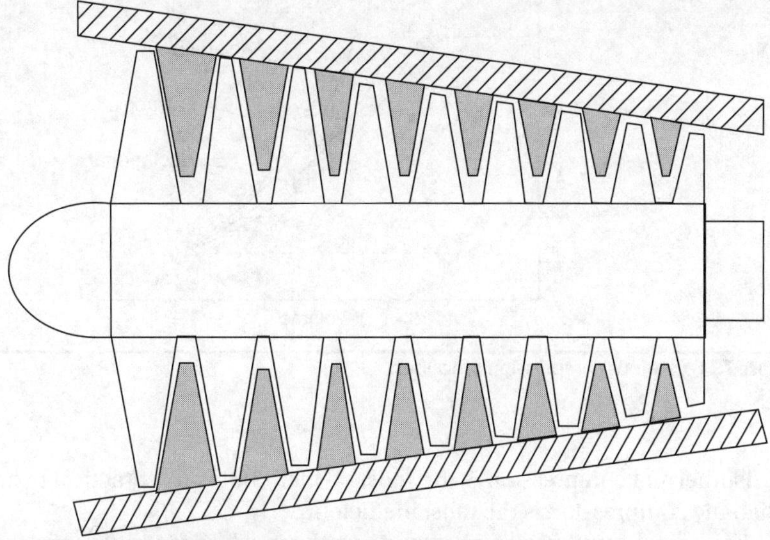

Figure 7.10 Cut-away of an axial-flow compressor.

7.6.6 Axial-Flow Compressors

Axial-flow compressors consist of a number of rows of rotating aerofoil blades (rotors) and stationary blades (stators) placed axially as shown in Fig. 7.10. The gas to be compressed passes parallel to the axis of rotation through each stage of rotors rotating at high speed, where velocity pressure is imparted. This is later converted to static pressure by diffusion when passing through the stators.

Axial compressors are used with gas turbines and in industrial large volume separation plants and blast furnaces.

7.7 Basic Theory of Compression

As air is compressed from a pressure P_1 to a higher pressure P_2, its volume reduces from V_1 to V_2. During compression, the temperature of the air also increases.

The compression process follows the path

$$PV^n = \text{constant} \tag{7.7}$$

As shown in Fig. 7.11, the path of compression can be

- "Isothermal," where the temperature is maintained constant (ideal case)
- "Adiabatic," where heat is not allowed to flow in or out during the compression
- "Polytropic," which is generally the case where heat exchange is allowed during compression but temperature is not maintained absolutely constant

In the expression $PV^n = \text{constant}$, the value of n is equal to 1 for isothermal, about 1.4 for adiabatic, and between about 1.2 to 1.3 for polytropic.

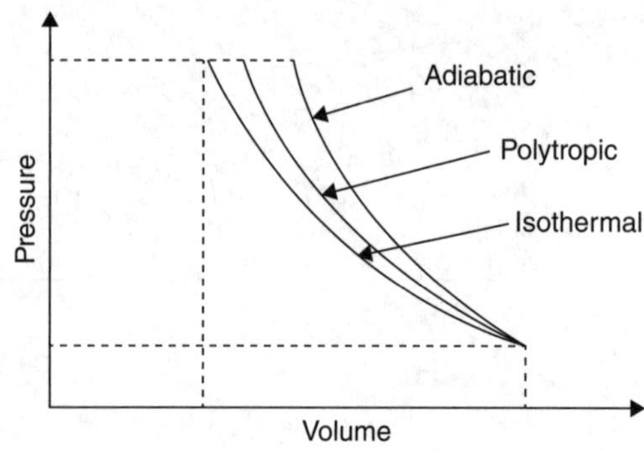

Figure 7.11 Various compression processes.

Isothermal compression is the most efficient but is not practical to achieve, while adiabatic compression is the most inefficient.

The actual compression process normally involves four basic stages: intake stage (where air enters the compression chamber), compression stage (where the air gets compressed), discharge stage (where the compressed air is discharged from the compressor), and expansion stage (where prior to the intake, the air trapped inside the compression chamber expands).

A typical compression process is shown in Fig. 7.12.

The work done in compression

$$W = \int V \, dP \tag{7.8}$$

The work done is represented by the area enclosed within the P–V diagram shown in Fig. 7.13.

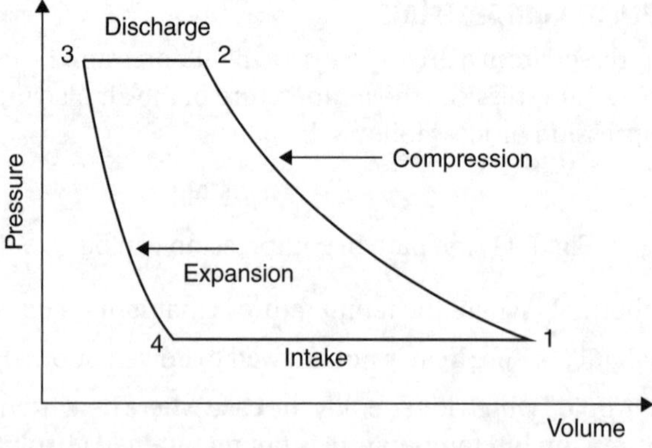

Figure 7.12 Typical compression process.

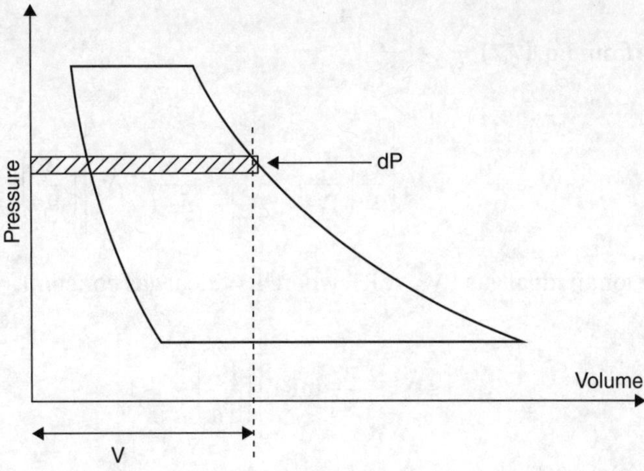

FIGURE 7.13 Work done in the compression process.

Using the relationship (7.7), the work done can be computed by performing the integration represented in Eq. (7.8) to provide the following two expressions for isothermal compression and polytropic or adiabatic compression:

$$\text{Work done } W = \int_1^2 V \; dP - \int_3^4 V \; dP$$

$$\text{Since } PV^n = C, \; V = C^{1/n} \times P^{-1/n}$$

$$\text{If the expansion process is neglected, } W = \int_1^2 V \; dP = \int_1^2 C^{1/n} \; P^{-1/n} \; dP$$

$$= \left[C^{1/n} \; \frac{P^{(1-1/n)}}{1-1/n} \right]_1^2$$

$$= \left[V \; P^{1/n} \; \frac{P^{(1-1/n)}}{1-1/n} \right]_1^2$$

$$= \left[\frac{V.P}{1-1/n} \right]_1^2 = \frac{n}{n-1} \, (P_2 V_2 - P_1 V_1)$$

$$= \frac{n}{n-1} P_1 V_1 \left(\frac{P_2 V_2}{P_1 V_1} - 1 \right)$$

Since from Eq. (7.7) $\dfrac{V_2}{V_1} = \left(\dfrac{P_2}{P_1}\right)^{-1/n}$

$$W = \frac{n}{n-1}\, P_1 V_1 \left[\frac{P_2}{P_1}\left(\frac{P_2}{P_1}\right)^{\frac{1}{n}} - 1\right] = \frac{n}{n-1}\, P_1 V_1 \left[\left(\frac{P_2}{P_1}\right)^{\frac{n-1}{n}} - 1\right]$$

Since for an ideal gas $PV = mRT$, where R = ideal gas constant,

$$W = \frac{n}{n-1}\, mRT_1 \left[\left(\frac{P_2}{P_1}\right)^{\frac{n-1}{n}} - 1\right] \tag{7.9}$$

For isothermal, since $n = 1$

$$W = \int_1^2 V\, dP = \int_1^2 CP^{-1}\, dP = PV\,[\ln P]_1^2$$

Therefore,

$$W = mRT_1 \ln\left(\frac{P_2}{P_1}\right) \tag{7.10}$$

where W = work done (J), m = mass of air being compressed (kg), R = specific gas constant for air (287 J/kg·K), T_1 = absolute temperature at the air compressor intake (K), and P_2/P_1 = compression ratio in absolute pressure values.

If the mass flow rate $\dot{m}$ is used instead of mass, m, the expressions for work done per second can be expressed as follows:

$$\dot{W} = \dot{m}RT_1 \ln\left(\frac{P_2}{P_1}\right), \text{ for isothermal compression} \tag{7.11}$$

$$\dot{W} = \frac{n}{n-1}\, \dot{m}RT_1 \left[\left(\frac{P_2}{P_1}\right)^{\frac{n-1}{n}} - 1\right], \text{ for polytropic and adiabatic compression} \tag{7.12}$$

where $\dot{W}$ = work done per second (W) and $\dot{m}$ = mass flow rate (kg/s).

7.8 Specific Power

Specific power for compressors is a measure of how much power is used to compress a specific quantity of air and is the ratio of power input to the compressor in kW to FAD in m³/s.

Example 7.4   Compute the mechanical power required to compress 1 m³/s of dry air at 30°C from 1 to 7.5 bar (both absolute pressures) for isothermal, polytropic (n = 1.3), and adiabatic (n = 1.4) compression. Take the density of air to be 1.2 kg/m³ and specific gas constant for air R to be 287 J/kg · K.

Solution

$$\dot{m} = 1\ \text{m}^3/\text{s} \times 1.2\ \text{kg/m}^3 = 1.2\ \text{kg/s}$$

Isothermal [from Eq. (7.11)]:

$$\dot{W} = 1.2 \times 287 \times (273 + 30) \times \ln\,(7.5/1)\ \text{W}$$

$$= 210\ \text{kW}$$

Adiabatic [from Eq. (7.12)]:

$$\dot{W} = 1.2 \times 287 \times (273 + 30) \times (1.4/0.4) \times [(7.5/1)^{0.4/1.4} - 1]\ \text{W}$$

$$= 284\ \text{kW}$$

Polytropic [from Eq. (7.12)]:

$$\dot{W} = 1.2 \times 287 \times (273 + 30) \times (1.3/0.3) \times [(7.5/1)^{0.3/1.3} - 1]\ \text{W}$$

$$= 267\ \text{kW}$$

Isothermal:

$$\text{Specific power} = 210\ \text{kW/1 m}^3/\text{s}$$

$$= 0.21\ \text{kW/(L/s)}$$

$$= 10\ \text{cfm/kW}$$

Adiabatic:

$$\text{Specific power} = 284\ \text{kW/1 m}^3/\text{s}$$

$$= 0.28\ \text{kW/(L/s)}$$

$$= 7.5\ \text{cfm/kW}$$

Polytropic:

$$\text{Specific power} = 267\ \text{kW/1 m}^3/\text{s}$$

$$= 0.27\ \text{kW/(L/s)}$$

$$= 8\ \text{cfm/kW}$$

(excluding motor and drive losses)   ▲

7.9   Efficiency

Isothermal efficiency is a measure of how close a particular process of compression is to the ideal case of isothermal compression. It is defined as the ratio of isothermal power to the actual power consumed.

$$\text{Isothermal efficiency} = \frac{\text{isothermal power}}{\text{actual power consumed}} \tag{7.13}$$

Example 7.5   Compute the isothermal efficiency of an air compressor operating at 1 m³/s of dry air at 30°C from 1 bar to 9 bar (both absolute pressures) and the measured power consumption is 325 kW. Take the density of air to be 1.2 kg/m³ and specific gas constant R to be 287 J/kg·K.

Solution

$$\dot{m} = 1 \text{ m}^3/\text{s} \times 1.2 \text{ kg/m}^3 = 1.2 \text{ kg/s}$$

Isothermal power [from Eq. (7.11)]:

$$\dot{W} = 1.2 \times 287 \times (273 + 30) \times \ln(9/1) \text{ W}$$
$$= 229.3 \text{ kW}$$

From Eq. (7.13):

$$\text{Isothermal efficiency} = 229.3/325 = 0.705 = 70.5\% \quad \blacktriangle$$

7.10   Multistage and Intercooling

The work required for compression depends on the compression ratio (ratio of discharge pressure to intake pressure) and the volume of air to be compressed. The amount of work required for compression is represented by the area enclosed in the P–V diagram (Fig. 7.14 showing pressure vs. volume) for the compression process.

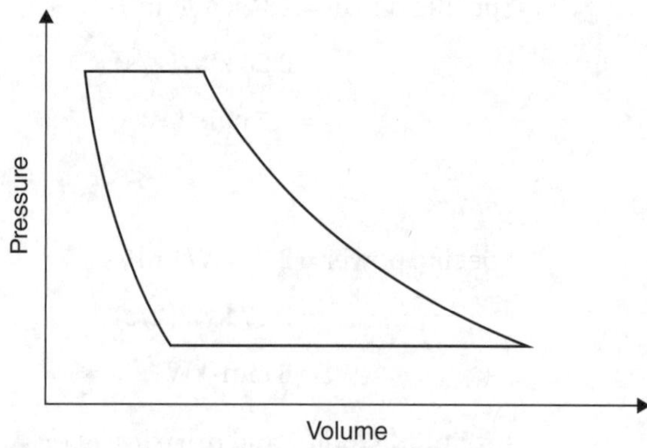

FIGURE 7.14   P–V diagram for single-stage compression.

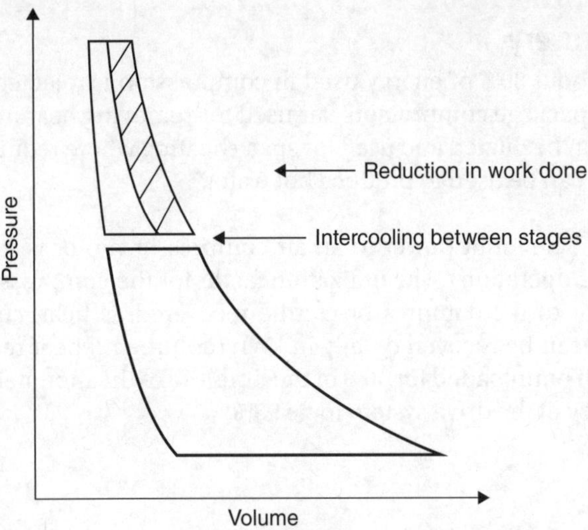

FIGURE 7.15   P–V diagram for two-stage compression.

In high-pressure applications, the compression can be divided into two or more stages where the air is compressed to an intermediate pressure and cooled before entering the next stage of compression. This helps to reduce the volume of air to be compressed in the subsequent stage. As shown in the P–V diagram for two-stage compression (Fig. 7.15), the work required, denoted by the area enclosed by the process, is reduced.

Example 7.6   FAD 0.05 m³/s of air is to be compressed from 1 bar (absolute) and 30°C to 36 bar (absolute). Compute the compressor power for single-stage and two-stage compression with intercooling, which reduces the temperature of the air to 30°C. Estimate the reduction in compressor power for two-stage compression. Take R to be 287 J/kg·K and density of air at 30°C to be 1.2 kg/m³. The compression pressure ratio for each stage is 1:6 and compression follows the path $PV^{1.3} = C$.

Solution

$$\dot{m} = 0.05 \text{ m}^3/\text{s} \times 1.2 \text{ kg/m}^3 = 0.06 \text{ kg/s}$$

Power for single-stage compression [using Eq. (7.12)]:

$$\dot{W} = 0.06 \times 287 \times (273 + 30) \times (1.3/0.3) \times [(36)^{0.3/1.3} - 1] \text{ W}$$

$$= 29 \text{ kW}$$

Power for two-stage compression:

$$\dot{W} = \dot{W}_1 + \dot{W}_2 = 2 \times \dot{W}_1 \text{ (since mass flow rate and inlet temperature are same for both stages)}$$

$$= 0.06 \times 287 \times (273 + 30) \times (1.3/0.3) \times [(6)^{0.3/1.3} - 1] \times 2 \text{ W}$$

$$= 23 \text{ kW}$$

Reduction in work done $= (29 - 23) = 6$ kW (20% reduction)   ▲

7.11 Heat Recovery

Generally about 80% of energy used in compression is rejected as heat. In air-cooled screw-type package compressors, air used for removing heat from the aftercooler and oil cooler can be ducted and used for space heating where required. Alternatively, heat exchangers can be used to produce hot water.

Example 7.7 The input power to an air compressor motor is measured to be 100 kW (during load operation). The utilization factor for the compressor is 60%. If 80% of the energy input to the compressor can be recovered as heat, compute the amount of energy that can be recovered daily in kWh (assume no heat recovery when the compressor is off or unloaded for 40% of the time). Take the motor efficiency to be 90% and the efficiency of the drive system to be 95%.

Solution

$$\text{Power at compressor} = 100 \times 0.9 \times 0.95 = 85.5 \text{ kW}$$

$$\text{Total energy input per day (load operation)} = 85.5 \times 0.6 \times 24 \text{ h} = 1231 \text{ kWh}$$

$$\text{Total amount of heat that can be recovered} = 1231 \times 0.8 = 985 \text{ kWh} \quad \blacktriangle$$

Example 7.8 FAD 0.05 m^3/s of air is to be compressed from 1 bar (absolute) and 30°C to 36 bar (absolute) in two stages of 1:6. Intercooling is provided between the two stages to cool the compressed air to 30°C. Compute the amount of heat rejected at the intercooler. Take R to be 287 J/kg·K, density of air at 30°C to be 1.2 kg/m^3, and specific heat capacity of air (C_p) to be 1.005 kJ/kg·K. The compression follows the path $PV^{1.3} = C$.

Solution
Mass flow rate of air, $\dot{m} = 0.05 \text{ m}^3/\text{s} \times 1.2 \text{ kg/m}^3 = 0.06 \text{ kg/s}$.
 From Eq. (7.1), for an ideal gas:

$$(P_1 \times V_1)/T_1 = (P_2 \times V_2)/T_2$$

Also from Eq. (7.7):

$$PV^n = C$$

Therefore, $P_1 \times V_1^{\,n} = P_2 \times V_2^{\,n}$

$$(T_2/T_1) = (P_2 \times V_2)/(P_1 \times V_1) = (P_2/P_1) \times (P_2/P_1)^{-1/n} = (P_2/P_1)^{1-1/n}$$

Therefore, $T_2 = (P_2/P_1)^{1-1/n} \times T_1 = (6)^{1-1/1.3} \times 303 = 458$ K (temperature of air leaving the first stage of the compressor, which is the temperature of air entering the intercooler).
 Heat rejected at the intercooler:

$$Q = m \times C_p\,(T_{in} - T_{out}) = 0.06 \times 1.005 \times (458 - 303) = 9.35 \text{ kW} \quad \blacktriangle$$

7.12 Dryers

Air contains moisture and when compressed, the same amount of moisture is then contained in a lower volume of air. This results in condensation, which needs to be removed as, otherwise, it could damage equipment, cause corrosion, and even affect the performance of systems.

Moisture therefore needs to be removed from compressed air before being supplied to most end users. The aftercooler that is used to cool the air after compression to bring it down to a usable temperature removes a significant amount of moisture from the air. In addition, dedicated dryers are used to remove further moisture from the compressed air.

The dryness of the air after the dryer is normally indicated as "dew point," which is the temperature to which the air needs to be cooled before the moisture in the air condenses. Therefore, air at a low dew point needs to be cooled to a lower temperature before moisture can condense, indicating that the moisture content of the air is low.

The dew point is pressure-dependent. The dew point of air can be stated at atmospheric pressure or at the operating pressure (pressure dew point). Therefore, if the mass of moisture in the air remains constant, the pressure dew point would be higher than the atmospheric dew point. For example, if the atmospheric dew point of a stream of air is −20°C, the respective pressure dew point at a pressure of 7 bar (gauge) will be +5°C. The relationship between atmospheric dew point and pressure dew point is listed in Table 7.3 for a few common system operating pressures.

The compressed air dew point (highest acceptable amount of moisture) is dependent on the process or application served by compressed air. Some of the typical dew point requirements are summarized in Table 7.4.

The required dew point dictates the type of dryer that needs to be used to achieve it. Generally, the lower the dew point, the higher will be the energy cost to operate the dryer. Table 7.5 provides a comparison of the achievable dew point temperatures and added energy consumption (to operate the dryer and to overcome pressure losses) for common types of dryers. The main types of dryers used in compressed air systems are described below.

Atmospheric dew point (°C)	Approximate pressure dew point at 3 bar (°C)	Approximate pressure dew point at 7 bar (°C)	Approximate pressure dew point at 15 bar (gauge) (°C)
+10	+32	+46	+57
+0	+20	+32	+43
−10	+6	+18	+29
−20	−5	+5	+15
−40	−27	−20	−13
−60	−48	−44	−37
−70	−60	−55	−49

TABLE 7.3 Pressure Dew Point Temperature Values

Application	Typical dew point at 7 bar (gauge) (°C)
Air motors	+7
Air agitation	−20
Machine tools	+7
Conveying granular materials	−20
Conveying powder materials	−40
Food and beverage industry	−70
Pharmaceutical industry	−70
Electronic industry	−70
Process control instruments	−20

TABLE 7.4 Pressure Dew Point Requirements

7.12.1 Refrigerant Dryers

Refrigerant dryers consist of an evaporator, condenser, compressor, and an expansion device and work based on the vapor compression cycle. Liquid refrigerant evaporates in the evaporator, which is then compressed and condensed back at a higher pressure in the condenser. Compressed air is passed over the evaporator coil, which cools the air to its dew point, resulting in condensation. An air-to-air heat exchanger is used to pre-cool the saturated air using the cool dry air after the evaporator. The typical arrangement of a refrigerant dryer is shown in Fig. 7.16.

Refrigerant dryers can normally achieve a dew point of about +2°C/3°C and are suitable for general applications. They generally have a low operating cost, low pressure drop (about 0.2 bar), and low maintenance cost as compared to desiccant dryers.

7.12.2 Desiccant Dryers

Desiccant dryers consist of desiccant materials that absorb moisture from the air and are able to achieve very low dew point temperatures, say, −70°C (image of a desiccant dryer is shown in Fig. 7.17). These dryers normally use two containers filled with desiccant and air is passed through one of the containers to produce dry air. Once the desiccant in this container is saturated with moisture absorbed from the stream of

Type of dryer	Achievable pressure dew point (°C)	Approximate added energy usage
Refrigerant dryers	0 to +3	5%
Heatless desiccant (air purging) dryers	−40 to −70	20%
Internally heated desiccant dryers	−40 to −70	18%
Externally heated desiccant dryers	−40 to −70	15%
Waste heat/heat of compression dryers	−40 to −70	10%

TABLE 7.5 Pressure Dew Point Requirements

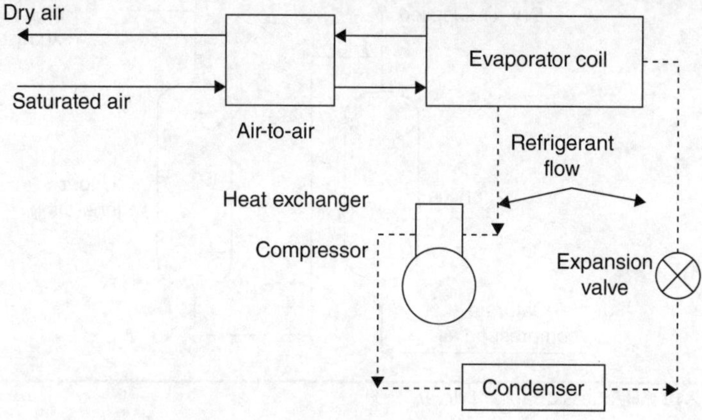

FIGURE 7.16   Arrangement of a refrigerant dryer.

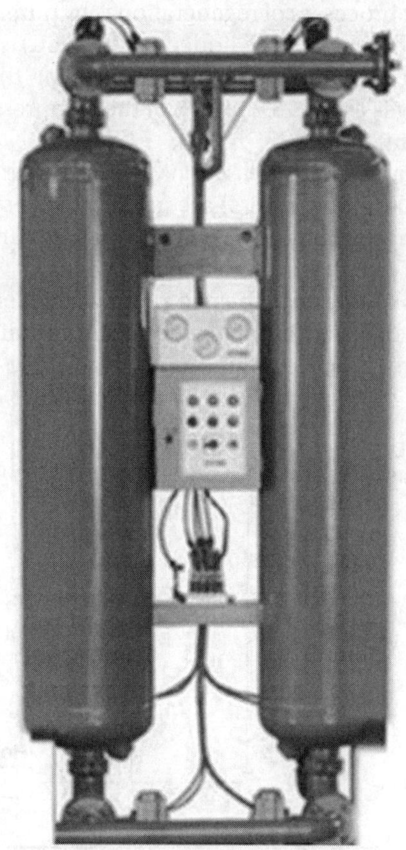

FIGURE 7.17   Image of a desiccant dryer.

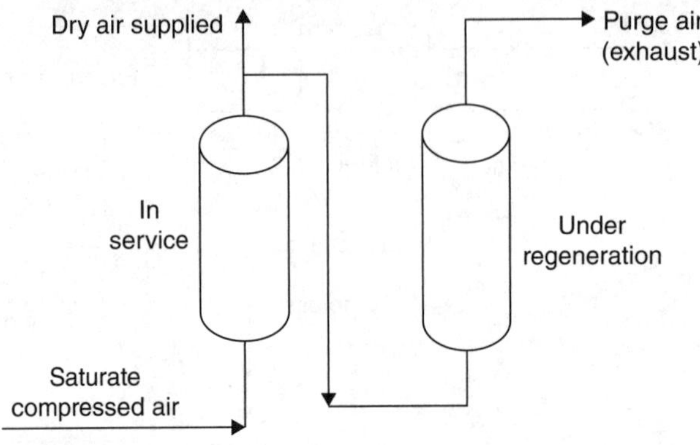

Figure 7.18   Heatless desiccant dryer.

compressed air, this container is taken out of service and the air is passed through the second container. The absorbed moisture is then removed from the desiccant in the container that is out of service by a "regeneration" process as shown in Fig. 7.18

One of the basic processes of regeneration is by purging where part of the dry compressed air (taken after the dryer cylinder in operation) is passed through the desiccant in the cylinder that is not in service and exhausting this air out. The amount of air required for purging is about 15% to 20% of the compressed air. Such dryers are therefore not very efficient.

To minimize purging compressed air losses, heaters embedded in the desiccant (internally heated desiccant dryers) are used in some systems where the heaters are turned on during regeneration. The purge air losses can be reduced to about half for internally heated dryers.

As an alternative, externally heated desiccant dryers with blowers can be used, where air is heated before entering the desiccant container being regenerated, which results in the need for much less purging air (Fig. 7.19).

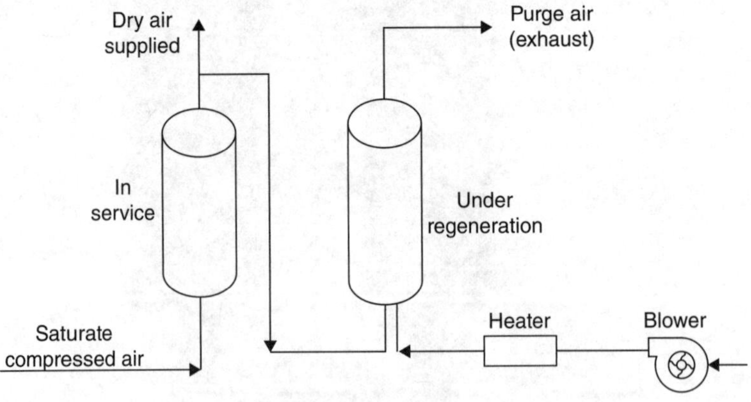

Figure 7.19   Desiccant dryer with an external blower and heater.

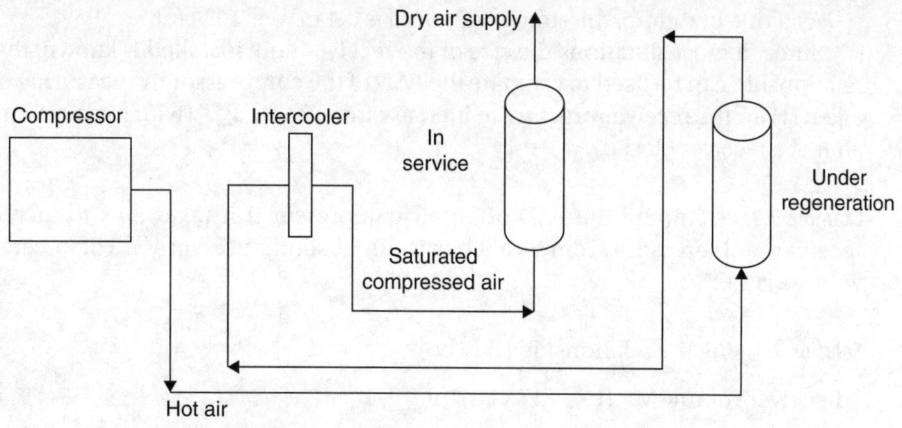

FIGURE 7.20 Heat-of-compression desiccant dryer.

Another method of regeneration is to use the heat produced by the compression process in heat-of-compression desiccant dryers. In such systems, part of the hot air after the compressor is passed through the desiccant to be regenerated before entering the aftercooler as shown in Fig. 7.20.

7.13 Receiver Tanks

In most compressed-air systems, the demand varies between a minimum and maximum value. Since compressors are selected based on a certain capacity, receiver tanks are used to act as storage devices so that the compressors can cycle between set load and unload pressures to provide supply air at a constant pressure to the users. Therefore, they help to minimize pressure fluctuations in the system and allow for sudden demand changes. They also help to remove condensate from the system.

The two main types of receivers used are (1) primary receivers installed near the air compressors between the aftercoolers and filters and (2) secondary receivers installed near loads with high intermittent loads.

The size (volume) of the required receiver depends on the maximum and minimum pressures set for the compressors to cut out and cut in the capacity (FAD) of the compressors and the time allowed for pressure to reduce from the maximum to minimum values and can be expressed as follows:

$$\text{Receiver volume } V = (t \times \text{FAD} \times \text{Pa})/(P_{max} - P_{min}) \qquad (7.14)$$

where V is in m^3, t = time required (s) to increase receiver pressure from P_{max} to P_{min}, P_{max} = maximum pressure at which the compressor cuts out (bar), P_{min} = minimum pressure at which the compressor cuts in (bar), Pa = atmospheric pressure (bar), and FAD = free air delivery required (m^3/s).

As a rule of thumb, the storage required is 0.4 m³ per 100 cfm.

Since in most situations the size of the receiver tank installed is known, the above relationship can be used to estimate the FAD of the compressor by measuring the time taken (t) for the receiver pressure to increase from P_{min} to P_{max} (with the discharge valve after the receiver closed).

Example 7.9   Compute the FAD of an air compressor if it takes 30 s to increase the receiver tank pressure from 7 to 9 bar (both absolute pressures). The receiver tank volume is 1 m³.

Solution   Using the relationship (7.14):

$$\text{Receiver volume } V = (t \times \text{FAD} \times \text{Pa})/(P_{max} - P_{min})$$

$$\text{Therefore, FAD} = V \times (P_{max} - P_{min})/(t \times \text{Pa}) = 1 \times (9-7)/30 \times 1 = 0.067 \text{ m}^3/\text{s} \quad \blacktriangle$$

7.14   System Pressure and Losses

The system pressure, which is the pressure at which compressed air is supplied by the compressors, depends on factors such as the pressure requirements of the end users, pressure losses in the system, storage capacity, and variation in demand.

Power consumed by compressors depends on the discharge pressure, as expressed by the relationship (7.11):

$$\dot{W} = \dot{m}RT_1 \left(\frac{n}{n-1}\right)\left[\left(\frac{P_2}{P_1}\right)^{\left(\frac{n-1}{n}\right)} - 1\right]$$

Therefore, a higher discharge pressure results in a higher compressor power (as the compressor has to work against a higher back pressure). The rule of thumb used in industry is a 7% to 8% increase in compressor power for every 1 bar increase in compressor discharge pressure.

Example 7.10   Compute the reduction in compressor power achievable when the system pressure is reduced from 9 to 7 bar (both absolute) for a compressor operating at 0.2 m³/s of air (FAD). Air intake temperature is 30°C, intake pressure is 1 bar (absolute), and the compression follows $PV^{1.3} = C$. Density of air can be taken as 1.2 kg/m³, while the specific gas constant R is 287 J/kg·K.

Solution   Using the relationship (7.12):
At 9 bar:

$$\dot{W} = 0.2 \times 1.2 \times 287 \times (273+30)\left(\frac{1.3}{0.3}\right)\left[\left(\frac{9}{1}\right)^{\left(\frac{0.3}{1.3}\right)} - 1\right]$$

$$= 59.7 \text{ kW}$$

$$\text{At 7 bar, } \dot{W} = 0.2 \times 1.2 \times 287 \times (273+30) \left(\frac{1.3}{0.3}\right)\left[\left(\frac{7}{1}\right)^{\left(\frac{0.3}{1.3}\right)} - 1\right]$$

$$= 51.3 \text{ kW}$$

Reduction in compressor power $= 59.7 - 51.3 = 8.4$ kW

(14% reduction in compressor power for a 2-bar reduction in pressure) ▲

Not only does a higher system pressure result in a higher compressor pressure, it also leads to increased system leaks and increased compressed air consumption by some end users (which, in turn, results in more compressor power).

Therefore, system pressure should be set at the minimum value required to satisfy equipment requirements. In systems where a single user requires a very much higher pressure than the other users, stand-alone compressors should be used for the system that requires a higher pressure so that the system serving the other users can operate at a lower pressure.

7.15   Compressed Air Leaks

Since compressed air is produced, stored, distributed, and used at a much higher pressure than the surrounding air, compressed air will leak through even the smallest openings, resulting in significant energy wastage. In many compressed-air systems, leaks account for about 20% to 30% of the total compressor output.

Common sources of leaks are pipe fittings, flexible tubes, couplings, pressure regulators, condensate traps, and pipe joints.

Compressed air leakage depends on the size of the opening, system pressure, and the shape of the opening (round or sharp).

Leakage rate can be estimated using Fliegner's formula for flow through an orifice:

$$m = \sqrt{(2/R)} \, CA \times P \times T^{-1/2} \tag{7.15}$$

where m is in kg/s, C is 0.65 for round holes, A is the orifice size in m^2, P is the pressure in Pa, and T is the absolute temperature in K.

However, it is not convenient to use the above relationship as the orifice size is normally not known. Therefore, the total leakage rate from a compressed-air system can be estimated by switching off all the equipment that uses compressed air, operating the compressor to charge the system to the operating pressure, and measuring the time taken for "load" and "unload" cycles of the compressor continuously for about 10 cycles.

The following expression can be used to estimate the system leakage rate:

$$\% \text{ of leakage} = T/(T+t) \times 100 \tag{7.16}$$

and

$$\text{System leakage } (m^3/\text{min}) = (Q \times T)/(T+t) \tag{7.17}$$

where Q = compressor capacity (FAD in m^3/min), T = average load time (min), and t = average unload time (min).

Example 7.11   A leak test carried out on a compressed air system with all the compressed air equipment switched off provided the following information:

$$\text{Average load time} = 3 \text{ min}$$

$$\text{Average unload time} = 10 \text{ min}$$

If the FAD of the compressor is 0.1 m^3/s, compute the percentage and quantity of leaks.

Taking the specific power for the compressor system to be 270 $kW/m^3/s$, estimate the energy wastage (in kWh/day), assuming the system operates 24 h a day.

Solution

$$\text{Load time} = 3 \text{ min}$$

$$\text{Unload time} = 10 \text{ min}$$

$$\text{FAD} = 0.1 \text{ m}^3/\text{s}$$

$$\% \text{ leakage} = [3/(3+10)] \times 100 = 23\%$$

$$\text{Leakage rate} = 0.1 \text{ m}^3/\text{s} \times 0.23 = 0.023 \text{ m}^3/\text{s}$$

$$\text{Power usage} = 0.023 \times 270 = 6.21 \text{ kW}$$

$$\text{Energy usage} = 6.21 \times 24 = 149 \text{ kWh/day} \quad \blacktriangle$$

7.16   Using Blowers for Tank Agitation

Often compressed air is used for tank agitation in applications such as plating to provide movement of the solution in the tank. In such situations, the use of compressed air can be replaced by installing blowers, which are much more energy efficient than air compressors.

The blowers should be selected to provide sufficient pressure and airflow rate. The minimum pressure requirement is dependent on the liquid level in tank height and can be estimated using Eq. (7.18).

$$\text{Minimum gauge pressure in bar} = [(\rho \times g \times h)/(1 \times 10^{-5})] + 0.05 \qquad (7.18)$$

where ρ = density of the fluid in the tank (kg/m^3), g = acceleration due to gravity (m/s^2), and h = height of the liquid in the tank (m).

$$\text{The airflow rate required for agitation in SCFM} = A \times F \qquad (7.19)$$

where A = tank surface area in m^2 and F = agitation factor between 10 and 20 in $SCFM/m^2$.

Example 7.12   A tank contains a solution with a density of 1200 kg/m^3. The liquid level in the tank is 1 m. The surface area of the tank is 2 m^2. It is currently being served by an air compressor having a specific power of 0.2 kW/SCFM.

Compute the required pressure and airflow rate for a suitable blower for this application, taking the agitation factor to be 15 SCFM/m². Estimate the reduction in power if the specific power for the blower is 0.08 kW/CFM.

Solution   From Eq. (7.18):

$$\text{Gauge pressure requirement} = [(1200 \times 9.81 \times 1)/(1 \times 10^{-5})] + 0.05 = 0.17 \text{ bar}$$

$$\text{The airflow rate required} = 2 \times 15 = 30 \text{ SCFM}$$

$$\text{Estimated reduction in power} = 30 \times (0.2 - 0.08) = 3.6 \text{ kW} \quad \blacktriangle$$

7.17   Intake Temperature

Compressor power consumption to compress a certain volume of air depends on the intake temperature of the air. If the intake air is hot, since the density of air is lower, the compressor needs to take in a larger volume of air to produce the same output.

Therefore, it is essential to provide adequate ventilation for compressor rooms to minimize the intake air temperature. Ventilation can be provided by natural means or by forced ventilation using fans and ducting systems.

Generally, a 5°C reduction in intake temperature of air results in a 1.5% reduction in compressor power.

Example 7.13   A compressor is located in a boiler room and draws air from the room at 40°C. The compressor power consumption is estimated to be 80 kW. Compute the savings in compressor power that can be achieved if outdoor air at 35°C can be supplied direct to the compressor.

Solution   From Eq. (7.12):

$$\dot{W} \propto T_1$$

Therefore, $(\dot{W}_2/\dot{W}_1) = (T_2/T_1)$

$$\dot{W}_2 = 80 \times [(273 + 35)/(273 + 40)] = 78.7 \text{ kW}$$

(approximately 1.5% reduction in kW for 5°C reduction in temperature) $\blacktriangle$

7.18   Controls

Controls in compressed-air systems are used to match the supply of compressed air to demand requirements. Since compressed air demand is normally not constant, good control is essential for efficient system performance and operation.

Controls need to be designed to optimize a system, utilizing strategies such as minimizing the operation of compressors, switching on compressors only when needed, ensuring the compressors in operation are fully loaded before switching on additional units, and switching on the most efficient unit based on the load.

Compressed-air systems are designed to provide varying amounts of compressed air to meet varying load requirements while maintaining a set system pressure. The system pressure is controlled within a band called the control range between a set maximum pressure and a minimum pressure.

Various control strategies are used to maintain the system pressure within the control range. The four most common compressor control strategies are listed below.

7.18.1   On/Off Control

This form of control involves switching on and off the compressor motor to control the amount of compressed air produced to meet the load. This type of control is used only for small compressors that have low duty cycles as, otherwise, frequent cycling will lead to overheating of the motor and excessive wear and tear of compressor parts.

7.18.2   Load/Unload Control

This control strategy uses two compressor operating modes: fully loaded and idling (unload). In the fully loaded mode of operation, the compressor runs at a fixed speed and operates at its rated capacity, while under the unload mode, the compressor and motor operate at the full-load speed, but the compressor does not produce compressed air. The main disadvantage of this mode of control is that power is consumed during unload operation. The power consumed during unload operation can be as high as 30% of the full-load power for a screw compressor. However, in the case of reciprocating compressors with multiple cylinders, the unload power can be relatively low as individual cylinders can be unloaded (for example, in a compressor with four cylinders, only one cylinder can be unloaded at 75% load).

7.18.3   Inlet Modulation

Inlet modulation control involves varying the volume of intake air to match load requirements. At part load, inlet airflow is reduced by vanes or modulating valves. This reduces inlet pressure and raises the compressor lift, leading to lower compressor efficiency. This mode is generally used for centrifugal compressors to modulate in the range of about 80% to 100% of the capacity.

7.18.4   VFD Control

This control strategy involves varying the capacity of compressors to match the load by varying the speed of the compressor using variable-frequency drives (VFDs). At full load, VFD compressors consume about 2% to 4% more power due to losses in the VFDs, but they can maintain good part-load efficiency up to about 50% (i.e., will consume about 50% of full-load power when operating at 50% load).

VFD control is usually used with rotary screw compressors. Due to the higher cost for screw compressors with VFDs and the higher power drawn at 100%, VFD compressors should normally be used to meet varying loads, while fixed-speed compressors are used to meet the base load.

Figure 7.21 provides a comparison of the variation in power consumed (as a percentage of the full-load power) for the various control strategies.

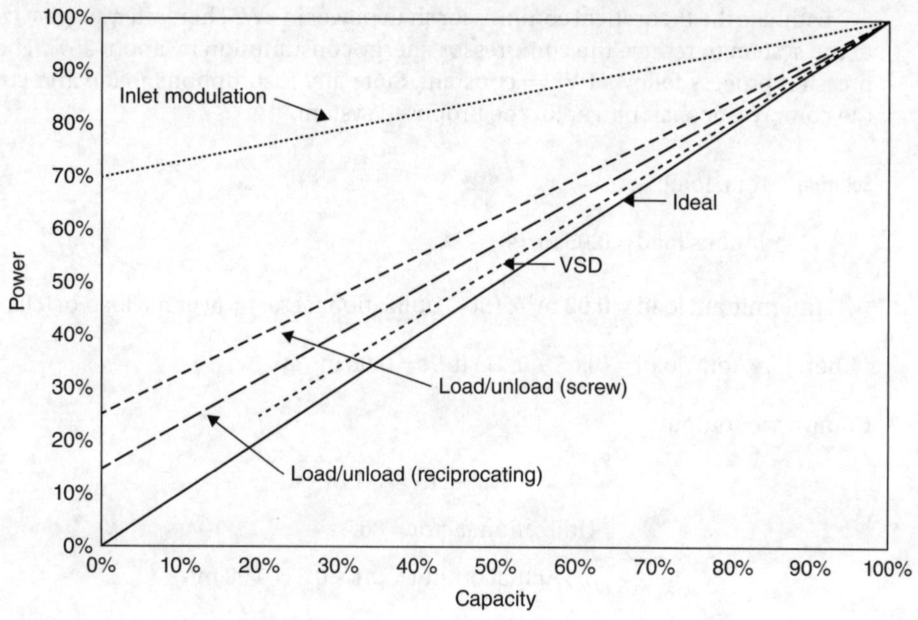

FIGURE 7.21 Power consumption for different control strategies.

Example 7.14 Consider the compressed-air system shown in Fig. 7.22. The system is installed in a boiler room where the air temperature is 40°C and operates 24 h a day to serve continuous and intermittent loads. The system operating pressures (values in absolute pressure) are shown in the figure.

The compressor has a rated capacity of 0.1 m³/s FAD. The compressor operates on load/unload mode and the utilization rate for the compressor is measured to be 80%. The compressor shaft power during unload operation is 30% of the shaft power during load operation.

The compressed air loads consist of two main types of loads: continuous loads of about 0.05 m³/s (FAD) and intermittent loads of 0.02 m³/s, which have a utilization factor of 50%. The dew point requirement for all the loads is −50°C.

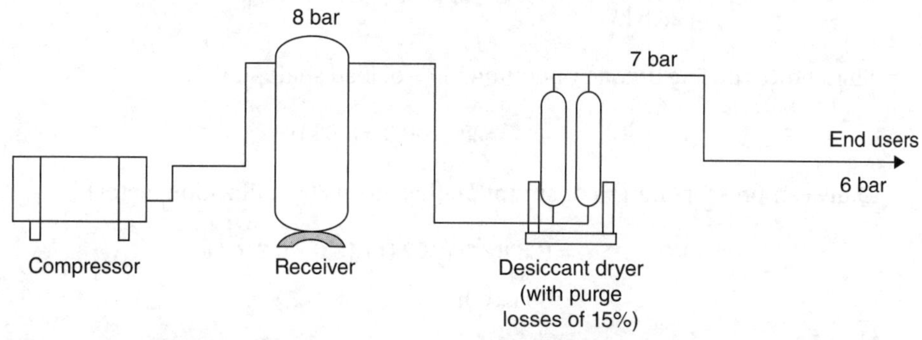

FIGURE 7.22 Diagram for Example 7.14.

Estimate the theoretical compressor shaft power in kW. Thereafter, suggest changes to the system to reduce the compressor energy consumption by about 25%. The compression process follows $PV^{1.3}$ = constant. State any assumptions made and compute the compressor shaft power for the proposed system.

Solution   Total load:

$$\text{Continuous load} = 0.05 \text{ m}^3/\text{s}$$

$$\text{Intermittent load} = 0.02 \text{ m}^3/\text{s} \ (50\% \text{ utilization factor}) = \text{average load of } 0.01 \text{ m}^3/\text{s}$$

$$\text{Therefore, total load} = (0.05 + 0.01) \text{ m}^3/\text{s} = 0.06 \text{ m}^3/\text{s}$$

Compressor output:

$$\text{Capacity} = 0.1 \text{ m}^3/\text{s}$$

$$\text{Utilization factor} = 80\%$$

$$\text{Actual output} = 0.1 \times 0.8 = 0.08 \text{ m}^3/\text{s}$$

Dryer purge losses:

$$\text{Assumption for dryer purge losses} = 15\% \text{ of compressed air production}$$

$$\text{Therefore, estimated purge air losses} = 0.15 \times 0.08 = 0.012 \text{ m}^3/\text{s}$$

Estimation of compressed air leaks:

$$\text{Approximate leakage rate} = \text{compressor output} - \text{purge losses} - \text{total load}$$

$$= (0.08 - 0.012 - 0.06) \text{ m}^3/\text{s}$$

$$= 0.008 \text{ m}^3/\text{s}$$

From Eq. (7.12), the estimated current shaft power (load condition)

$$W = 0.1 \times 1.2 \times 287 \times (273 + 40) \times (1.3/0.3) \times [(8/1)^{0.3/1.3} - 1] \text{ W}$$

$$= 28.8 \text{ kW}$$

Shaft power during unload operation (30% of load shaft power)

$$W = 28.8 \times 0.3 = 8.6 \text{ kW}$$

Daily compressor energy consumption (based on 80% utilization factor)

$$= (28.8 \times 0.8 \times 24) + (8.6 \times 0.2 \times 24)$$

$$= 594 \text{ kWh}$$

Some of the possible improvement measures that can be taken to reduce the compressor energy consumption are as follows:

- Reducing the intake air temperature to 30°C by using a ducted inlet to compressor, assuming that the outdoor temperature is 30°C
- Using an externally heated dryer to reduce purge losses
- Eliminating leaks
- Reducing system losses by about 0.5 bar (assuming that some of the distribution pipes can be resized and by installing an additional storage tank closer to the intermittent loads

Estimated shaft power for system after modification (load condition)

$$W = 0.1 \times 1.2 \times 287 \times (273 + 30) \times (1.3/0.3) \times [(7.5/1)^{0.3/1.3} - 1] \text{ W}$$

$$= 26.7 \text{ kW}$$

During unload operation (assuming shaft power will be 30% of load value)

$$W = 26.7 \times 0.3 = 8 \text{ kW}$$

Once leaks and purge air are eliminated, the total compressed air requirement will be 0.06 m³/s.

Therefore, the new utilization factor will be 0.06/0.1 = 0.6 (load divided by compressor capacity).

Daily compressor energy consumption (based on the new utilization factor of 60%)

$$= (26.7 \times 0.6 \times 24) + (8 \times 0.4 \times 24)$$

$$= 461 \text{ kWh}$$

$$\text{Energy savings} = (594 - 461)/594 = 22\%$$

The above energy savings estimation is based on using the same compressor. However, if a smaller compressor is installed or if the capacity of the existing compressor can be reduced using a VFD, the shaft power would be

$$W = (0.06) \times 1.2 \times 287 \times (273 + 30) \times (1.3/0.3) \times [(7.5/1)^{0.3/1.3} - 1] \text{ W}$$

$$= 16 \text{ kW}$$

Daily energy consumption

$$= 16 \times 24$$

$$= 384 \text{ kWh}$$

Therefore, the energy savings $= (594 - 384)/594 = 35\%$ ▲

CHAPTER 8

Heat Recovery Systems

8.1 Introduction

Many industrial processes require heating of various fluids, which is achieved by equipment such as steam heat exchangers, combustion heaters, and electric heaters. All these heating processes require energy input. In the majority of cases, the heated fluids need to be cooled to lower temperatures after completion of the heating process. Cooling of a substance involves removal of heat, which in many instances requires energy. This requirement for heating and cooling is often simultaneous in many industrial processes. Therefore, if the heat removed from one fluid being cooled can be used to heat another fluid that requires the temperature to be increased, the overall energy efficiency of the manufacturing process can be improved.

Waste heat is available from sources such as furnaces, gas turbines, boilers, air compressors, and various manufacturing processes and can be used to heat process fluids, as shown in Fig. 8.1.

A very common application of heat recovery is in boilers. The exhaust from boilers is at a considerably high temperature and contains a significant amount of the input energy. This waste heat can be recovered using a heat recovery heat exchanger (economizer) and utilized to pre-heat the boiler feedwater (Fig. 8.2), resulting in less fuel usage by the boiler.

8.2 Modes of Heat Transfer

8.2.1 Conduction

Heat transfer by conduction takes place when there is a temperature gradient across a solid object (Fig. 8.3). The rate of heat transfer depends on the thermal conductivity of the material, its thickness, the temperature gradient, and the surface area available for heat transfer.

The rate of heat transfer by conduction can be expressed using Fourier's law as follows:

$$Q_{cond} = kA\frac{dT}{dx} \tag{8.1}$$

where k = thermal conductivity of the material, A = area (perpendicular to heat flow), dT = temperature gradient $(T_1 - T_2)$, and dx = thickness.

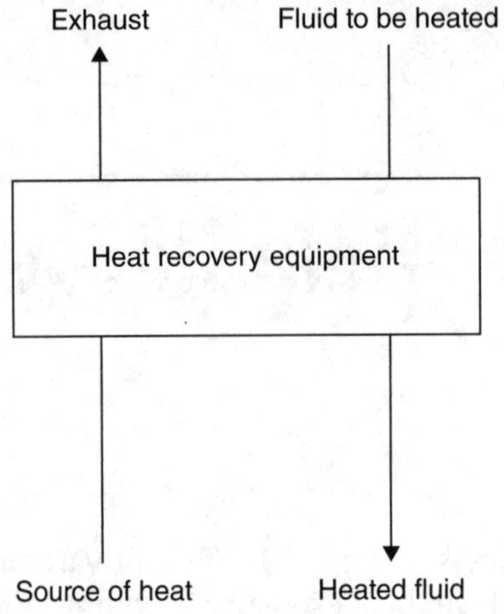

FIGURE 8.1   Arrangement of a heat recovery system.

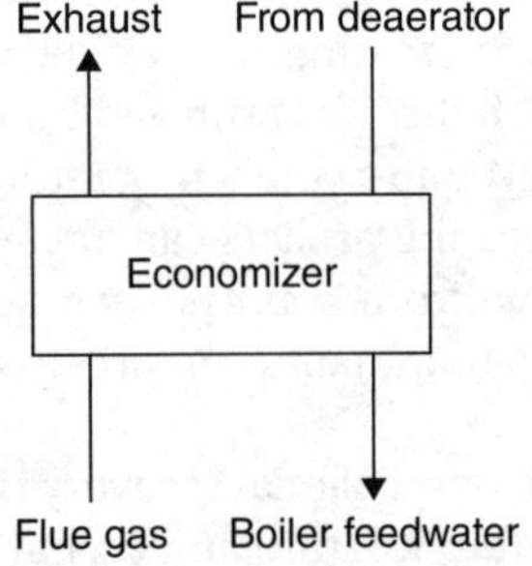

FIGURE 8.2   Typical heat recovery system for boiler exhaust.

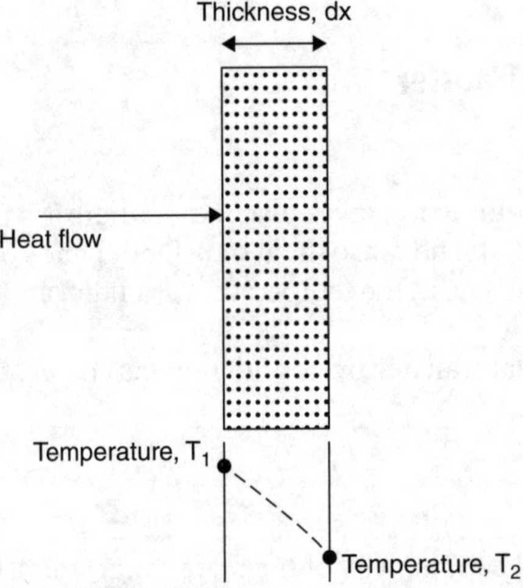

FIGURE 8.3   Heat transfer by conduction.

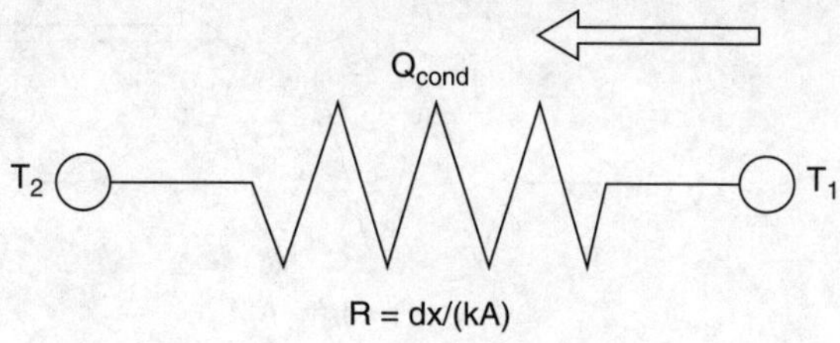

FIGURE 8.4 Thermal analogy of Ohm's law.

The above can be compared with Ohm's law, where potential difference (V) = current (I) × resistance (R). The thermal analogy (Fig. 8.4) would be $(T_1 - T_2) = dT = Q \times R$. Therefore, the thermal resistance $R = dx/kA$.

8.2.2 Convection

Heat transfer by convection takes place when a fluid comes into contact with a surface at a different temperature (Fig. 8.5). Heat transfer by convection can take place at both the inner and outer surfaces of an object.

Heat transfer by convection can be expressed using Newton's law of cooling, as follows:

$$Q_{conv} = h_c \, A \, (T_s - T_f) \tag{8.2}$$

where h_c = surface heat transfer coefficient, A = surface area, T_s = surface temperature, and T_f = fluid temperature.

The amount of convective heat transfer depends on the surface area, temperature difference between the surface and fluid, and the surface heat transfer coefficient. The

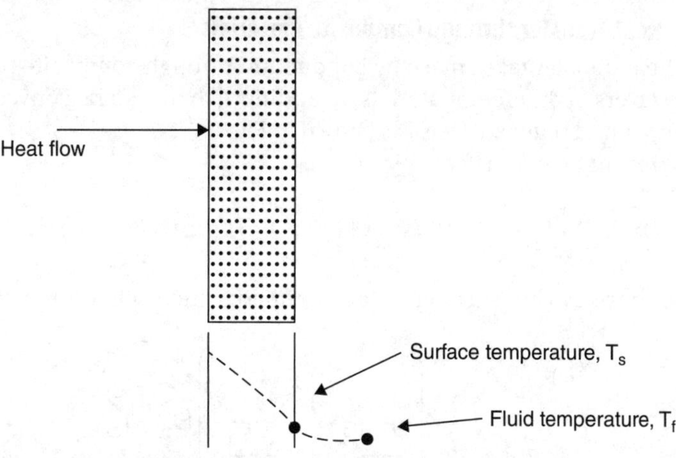

FIGURE 8.5 Convective heat transfer at the inner surface.

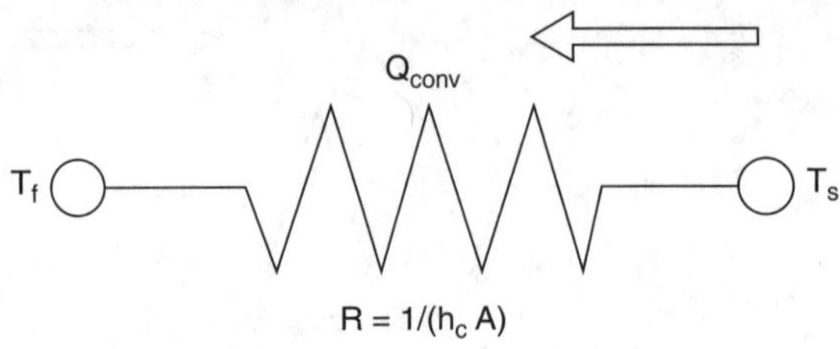

FIGURE 8.6   Ohm's law analogy applied to convective heat transfer.

surface heat transfer coefficient is dependent mainly on the velocity of the fluid motion over the solid surface. Ohm's law analogy is shown in Fig. 8.6.

8.2.3   Radiation

Heat transfer by radiation takes place due to electromagnetic waves. The amount of radiant heat transmission between two surfaces depends on the absolute surface temperatures of the bodies exchanging heat and the area of the body at a higher temperature. Radiant heat transfer can be expressed as follows:

$$Q_{rad} = \sigma \varepsilon_1 A_1 (T_1^4 - T_2^4) \tag{8.3}$$

where σ = Stefan–Boltzmann constant, A_1 = area of surface 1, ε_1 = emissivity of surface 1, T_1 = absolute temperature of surface 1, and T_2 = absolute temperature of surface 2.

Radiant heat transfer can also be described by a simple expression using the radiant heat transfer coefficient (h_r), as follows:

$$Q_{rad} = h_r A_1 (T_1 - T_2) \tag{8.4}$$

8.2.4   Heat Transfer through Composite Materials

Often, heat transfer takes place by conduction through composite materials made up of a few layers of different materials. In addition, there can be convective heat transfer at the inner and outer surfaces, as shown in Fig. 8.7.

The rate of heat transfer

$$Q = h_A A (T_A - T_1) = (k_1/x_1) A (T_1 - T_2) = (k_2/x_2) A (T_2 - T_3) = h_B A (T_3 - T_B)$$

The above can be rearranged and summed, which will result in the following:

$$\frac{Q}{A} = \left(\frac{1}{\dfrac{1}{h_A} + \dfrac{x_1}{k_1} + \dfrac{x_2}{k_2} + \dfrac{1}{h_B}} \right) (T_A - T_B) \tag{8.5}$$

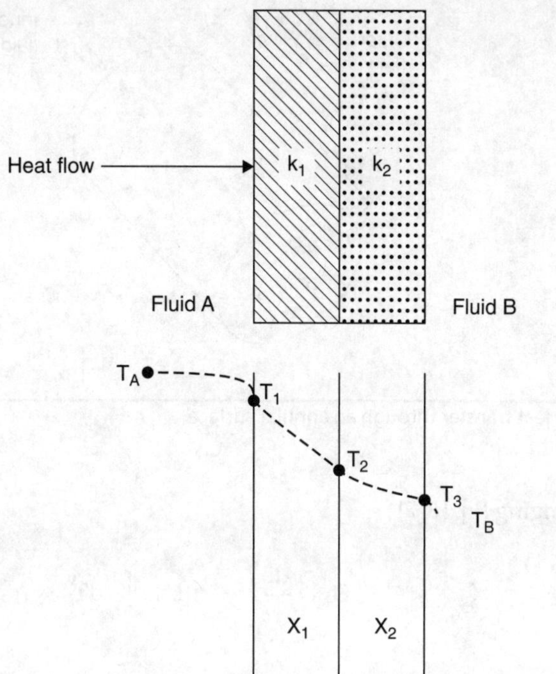

FIGURE 8.7   Heat transfer across a composite material.

Equation (8.5) can be rewritten as

$$Q = U\,A\,\Delta T$$

where the overall heat transfer coefficient

$$U = \left(\cfrac{1}{\cfrac{1}{h_A} + \cfrac{x_1}{k_1} + \cfrac{x_2}{k_2} + \cfrac{1}{h_B}} \right) \tag{8.6}$$

8.2.5   Radial Heat Transfer

Often, heat transfer between two fluids takes place through an annular surface such as across a pipe surface, as shown in Fig. 8.8.

Equation (8.1) can be expressed as

$$Q_{cond} = K\,A\,\frac{dT}{dr} \tag{8.7}$$

where k = thermal conductivity of the material, A = area (perpendicular to heat flow) = $2\pi r L$, dT = temperature gradient ($T_1 - T_2$), and dr = thickness ($r_2 - r_1$).

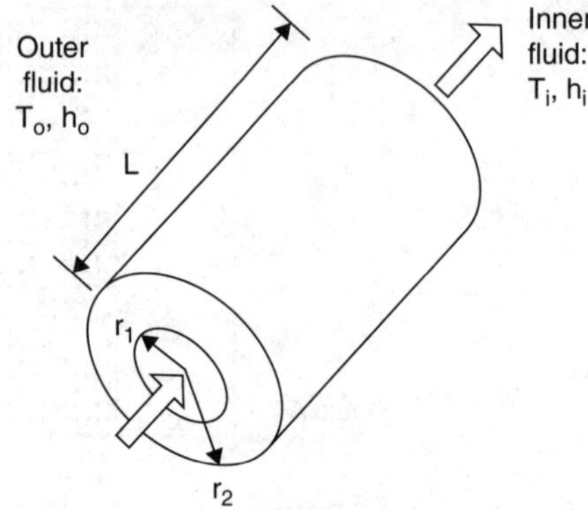

Figure 8.8   Heat transfer through an annular surface.

Rearranging Eq. (8.7)

$$Q \int_{r_1}^{r_2} \frac{dr}{r} = -2\pi \, k \, L \int_{T_1}^{T_2} dT \tag{8.8}$$

Therefore

$$Q = \frac{2\pi \, k \, L \, (T_1 - T_2)}{\ln \dfrac{r_2}{r_1}}$$

Heat flow at the inner and outer surfaces by convection:

$$Q_i = h_i \, 2\pi r_1 \, L \, (T_i - T_1)$$

$$Q_o = h_o \, 2\pi r_2 \, L \, (T_2 - T_o)$$

Since, $Q = Q_i = Q_o$

$$\frac{Q}{L} = \left(\frac{1}{\dfrac{1}{2\pi r_1 h_i} + \dfrac{\ln \dfrac{r_2}{r_1}}{2\pi k_1} + \dfrac{1}{2\pi r_2 h_o}} \right) (T_i - T_o) \tag{8.9}$$

8.3 Heat Exchangers

Heat exchangers can be classified according to the flow arrangement of the two fluids exchanging heat and are described in the following sections.

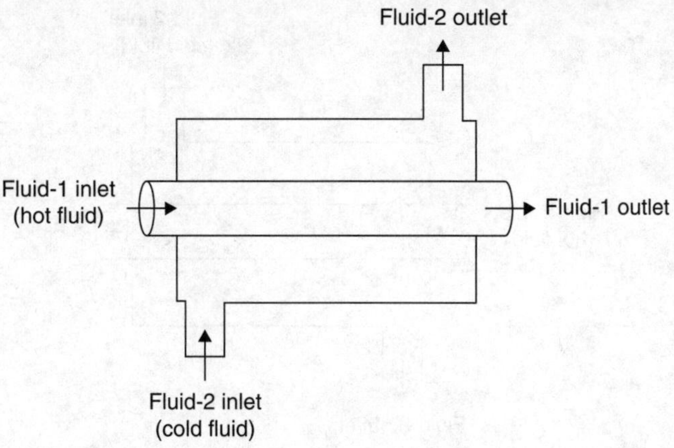

FIGURE 8.9   Arrangement of a parallel-flow heat exchanger.

8.3.1   Parallel-Flow Heat Exchangers

In parallel-flow heat exchangers, both the fluids enter the heat exchanger at the same end and flow parallel to each other before being discharged at the opposite end. The arrangement and temperature profile for a concentric tube (double pipe) parallel-flow heat exchanger are shown in Figs. 8.9 and 8.10. One main disadvantage of this type of heat exchanger is that the temperature difference between the two fluids, which is the driving potential for heat transfer, reduces with distance from the inlet to the outlet.

8.3.2   Counterflow Heat Exchangers

In counterflow devices, the two fluids enter at the two opposite ends of the heat exchanger and, as a result, are able to maintain a relatively better temperature difference between the two fluids when compared to parallel-flow exchangers. Therefore,

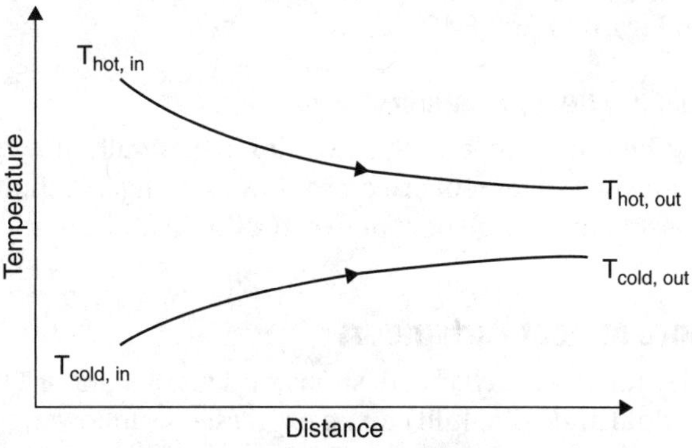

FIGURE 8.10   Temperature profile for a parallel-flow heat exchanger.

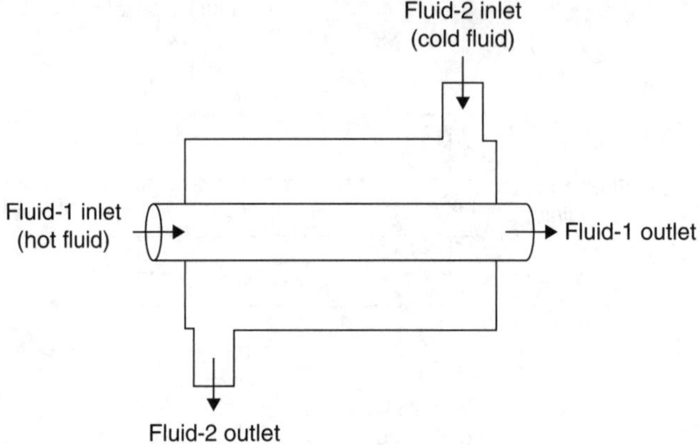

FIGURE 8.11 Arrangement of a counterflow heat exchanger.

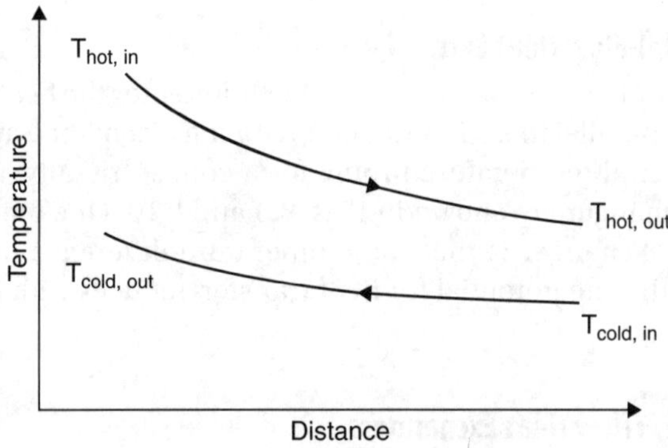

FIGURE 8.12 Temperature profile for a counterflow heat exchanger.

for the same heat transfer capacity, a counterflow heat exchanger will require a shorter length than a parallel-type one.

The typical arrangement and temperature profile for a counterflow heat exchanger are shown in Figs. 8.11 and 8.12.

8.3.3 Cross-Flow Heat Exchangers

In cross-flow heat exchangers, the two fluids generally flow perpendicular to each other. The typical arrangement of a cross-flow exchanger is shown in Fig. 8.13. They are commonly used to exchange heat between a liquid and a gas.

8.4 Performance of Heat Exchangers

If we consider the heat exchangers shown in Figs. 8.9 and 8.11, the heat-transfer rate from the hot fluid and cold fluid can be expressed as follows:

$$Q = m_{hot} \times C_{P_{hot}} \times (T_{hot\,in} - T_{hot\,out}) = m_{cold} \times C_{P_{cold}} \times (T_{cold\,out} - T_{cold\,in}) \qquad (8.10)$$

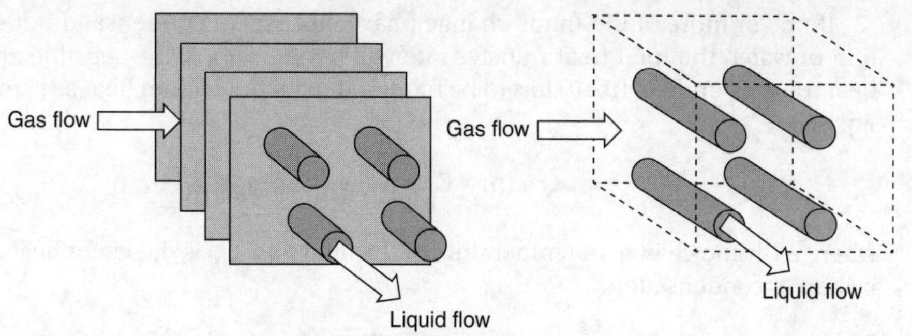

Figure 8.13 Arrangement of a cross-flow heat exchanger.

Material	Specific heat capacity (J/kg · K)
Water (27°C)	4180
Air (27°C)	1007
Engine oil (27°C)	1909
Carbon steel	434
Stainless steel	480
Copper	385

Table 8.1 Specific Heat Capacities of Common Fluids

where m_{hot} and m_{cold} are the mass flow rates of the hot and cold fluids, Cp_{hot} and Cp_{cold} are the specific heat capacities of the hot and cold fluids (see Table 8.1), $T_{hot\,in}$ and $T_{hot\,out}$ are the inlet and outlet temperatures of the hot fluid, and $T_{cold\,in}$ and $T_{cold\,out}$ are the inlet and outlet temperatures of the cold fluid, respectively.

Example 8.1 In a water-to-water heat exchanger, hot water from an industrial cooling process is used to heat cold water used in another process. The hot water enters the heat exchanger at 90°C and exits it at 45°C. If the hot water flow rate is 5 kg/s, compute the total amount of heat transferred from the hot water to the cold water. If the cold water flow rate is also 5 kg/s and enters the heat exchanger at 30°C, estimate the temperature of this stream of water at the exit of the heat exchanger. Take the specific heat capacity of water to be 4.19 kJ/kg · K.

Solution Using Eq. (8.10):

$$Q = m_{hot} \times Cp_{hot} \times (T_{hot\,in} - T_{hot\,out})$$

$$Q = 5 \times 4190 \times (90 - 45) = 942.75 \text{ kW}$$

$$\text{Also, } Q = m_{cold} \times Cp_{cold} \times (T_{cold\,out} - T_{cold\,in})$$

$$\text{Therefore, } 942{,}750 = 5 \times 4190 \times (T_{cold\,out} - 30)$$

$$T_{cold\,out} = 75°C \quad \blacktriangle$$

If one or more of the fluids change phase, like steam condensation or evaporation of water, the total heat transfer rate will be the sum of the sensible and latent heat transfer, and Eq. (8.10) has to be modified to include latent heat as expressed in Eq. (8.11):

$$Q = (m \times C_p \times \Delta T) + (m \times h_{fg}) \tag{8.11}$$

where ΔT is the change in temperature of the fluid and h_{fg} is the latent heat of vaporization or condensation.

Example 8.2   Exhaust flue gas from a furnace is used to heat water using a heat recovery heat exchanger. Water enters the heat recovery unit at a temperature of 45°C and the flow rate is 1 m³/h. Compute the heat recovery rate for the following two cases:

- Water exits the heat exchanger in liquid state at 100°C.
- Water exits the heat exchanger as saturated vapor (low pressure steam) at 100°C.

Take the specific heat capacity of water to be 4.195 kJ/kg·K, density of water to be 1000 kg/m³, and the latent heat of vaporization of water at 100°C to be 2257 kJ/kg.

Solution

(1) Using Eq. (8.10):

$$Q = m_{cold} \times Cp_{cold} \times (T_{cold\,out} - T_{cold\,in})$$

$$Q = (1000/3600) \times 4.195 \times (100 - 45) = 64.1 \text{ kW}$$

(2) Using Eq. (8.11):

$$Q = (m \times Cp \times \Delta T) + (m \times h_{fg})$$

$$Q = [(1000/3600) \times 4.195 \times (100 - 45)] + [(1000/3600) \times 2257] = 691 \text{ kW}$$

(Note that the heat-transfer rate with phase change is many times the value of pure sensible heat transfer.)   ▲

The total heat transfer rate can also be expressed as

$$Q = U \times A \times \Delta T_m \tag{8.12}$$

where U is the overall heat transfer coefficient for the heat exchanger, A is the heat transfer area, and ΔT_m is the mean temperature difference between the fluids.

The temperature difference between the two fluids (ΔT) varies with position and can be expressed as a mean value using the log mean temperature difference (LMTD). The LMTD is a logarithmic average of the temperature differences between the two fluids at each end of the heat exchanger.

$$\text{LMTD} = \frac{\Delta T_2 - \Delta T_1}{\ln\left(\dfrac{\Delta T_2}{\Delta T_2}\right)} \tag{8.13}$$

where ΔT_2 = temperature difference between the fluids at one end of the heat exchanger and ΔT_1 = temperature difference between the fluids at the other end of the heat exchanger.

Equation (8.12) can be written as

$$Q = U \times A \times \text{LMTD} \tag{8.14}$$

The above relationship applies to both parallel-flow and counterflow heat exchangers.

Example 8.3   A double-pipe counterflow heat exchanger is used to heat water from 35 to 90°C. The water mass flow rate is 1 kg/s. The water stream is heated using hot oil from a process at 140°C with a mass flow rate of 2 kg/s. The inner diameter of the heat exchanger pipe is 100 mm. The overall heat transfer coefficient of the heat exchanger (U) is 500 W/m·K and the specific heat capacities of water and oil are 4.19 kJ/kg·K and 1.9 kJ/kg·K. Compute the LMTD and length of the heat exchanger needed to achieve the required heating.

Solution   From Eq. (8.10):

$$Q = m_{cold} \times Cp_{cold} \times (T_{cold\ out} - T_{cold\ in})$$

$$Q = 1 \times 4.19 \times (90 - 35) = 230 \text{ kW}$$

$$\text{Also, } Q = m_{hot} \times Cp_{hot} \times (T_{hot\ in} - T_{hot\ out})$$

$$\text{Therefore, } Q = 230 = 2 \times 1.9 \times (140 - T_{hot\ out})$$

$$T_{hot\ out} = 79.5°C$$

From equation (8.13):

$$\text{LMTD} = \frac{\Delta T_2 - \Delta T_1}{\ln\left(\dfrac{\Delta T_2}{\Delta T_2}\right)}$$

$$\Delta T_2 = (140 - 90) = 50°C$$

$$\Delta T_1 = (79.5 - 35) = 44.5°C$$

$$\text{LMTD} = (50 - 44.5)/\ln{(50/44.5)} = 47.2°\text{C}$$

$$Q = U \times A \times \text{LMTD}$$

$$230{,}000 = 500 \times A \times 47.2$$

Therefore, the surface area for heat transfer, $A = 9.75 \text{ m}^2$

$$\text{Also, } A = \pi \times D \times \text{Length}$$

$$\text{Therefore, } 9.75 = 3.14 \times 0.1 \times L$$

The required heat exchanger length, $L = 31$ m   ▲

In the case of crossflow heat exchangers where one fluid is at a fixed temperature at all locations (heat sink), the LMTD can be expressed using the LMTD for a counterflow heat exchanger multiplied by a correction factor where

$$\text{LMTD}_{\text{cross-flow}} = F \times \text{LMTD}_{\text{counterflow}} \tag{8.15}$$

Correction factors "F" are available in charts for various configurations.

The correction factors when both fluids are unmixed (e.g. car radiator) and when one is unmixed and the other is mixed are shown in Figs. 8.14 and 8.15.

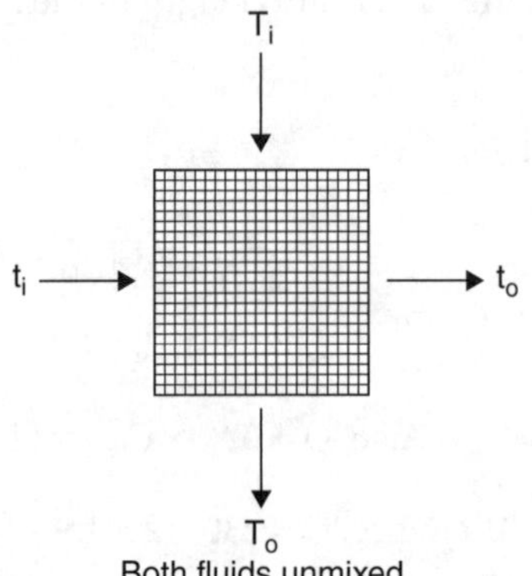

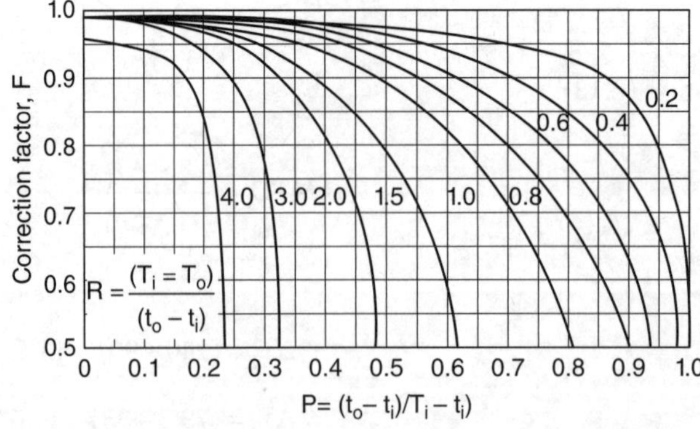

FIGURE 8.14   Correction factor F when both fluids are unmixed.

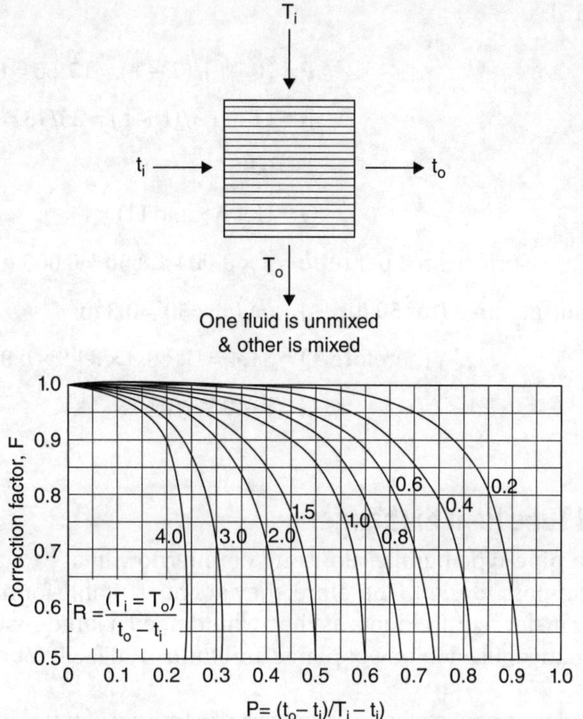

FIGURE 8.15   Correction factor F when one fluid is unmixed and the other is mixed.

Example 8.4   In a radiator that is a crossflow water-to-air heat exchanger where both fluids are unmixed, hot water enters the tubes at 80°C at a rate of 0.3 kg/s and leaves at 55°C. The radiator has 50 tubes of internal diameter 0.4 cm and length 50 cm in a closely spaced plate-finned matrix. Air flows across the radiator through the fin spaces and is heated from 25 to 40°C. Determine the overall heat transfer coefficient U of this radiator based on the inner surface area of the tubes. Take the specific heat capacity of water as 4190 J/kg·K.

Solution

$$Q = m \times C_p \times \Delta T = 0.3 \times 4.190 \times (80 - 55) = 31.4 \text{ kW}$$

$$LMTD = \frac{\Delta T_2 - \Delta T_1}{\ln\left(\dfrac{\Delta T_2}{\Delta T_2}\right)}$$

$$\Delta T_2 = (80 - 40) = 40°C$$

$$\Delta T_1 = (55 - 25) = 30°C$$

$$LMTD = (40 - 30)/\ln(40/30) = 34.8°C$$

$$LMTD_{crossflow} = F \times LMTD_{counterflow}$$

For Fig. (8.15):

$$P = (t_o - t_i)/(T_i - t_i) = 15/55 = 0.27$$

$$R = (T_i - T_o)/(t_o - t_i) = 25/15 = 1.7$$

$$F = 0.96$$

$$Q = U \times A \times LMTD \times F$$

$$\text{Surface area per tube} = \pi \times 0.004 \times 0.50 = 0.006 \text{ m}^2$$

$$\text{Total surface area (for 50 tubes)} = 0.006 \times 50 = 0.3 \text{ m}^2$$

$$\text{Therefore, } Q = 31.4 = U \times 0.3 \times 34.8 \times 0.96$$

$$U = 3140 \text{ W/m}^2 \cdot \text{K} \quad \blacktriangle$$

8.5   Shell and Tube Heat Exchangers

The double-pipe-type parallel-flow and counterflow heat exchangers (Figs. 8.9–8.11) are of "single-pass" design. Therefore, they are not suitable for many applications with high heat transfer requirements as they would need a large space. An alternative type of heat exchanger used in power plants and industrial facilities are shell and tube heat exchangers.

Shell and tube heat exchangers consist of a bundle of tubes inside a large shell. One of the fluids to be heated or cooled flows through the tubes, while the other fluid that either provides heat or absorbs heat flows over the tubes inside the shell. The fluids exchanging heat can be in liquid form where only sensible heat transfer takes place or can change phase for latent heat transfer as is the case in power plants where steam is condensed to liquid in the shell of the heat exchanger.

Shell and tube heat exchangers can be classified according to the number of "passes" the fluid flows through the shell or tube arrangement. Figures 8.16–8.18 show the arrangement of single- and two-pass heat exchangers.

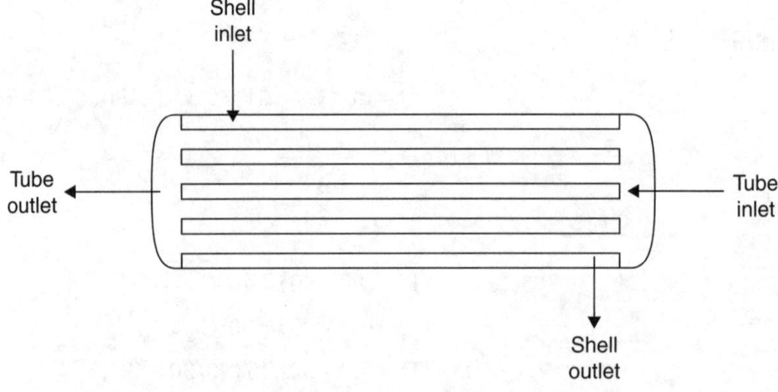

FIGURE 8.16   One-shell pass and one-tube pass.

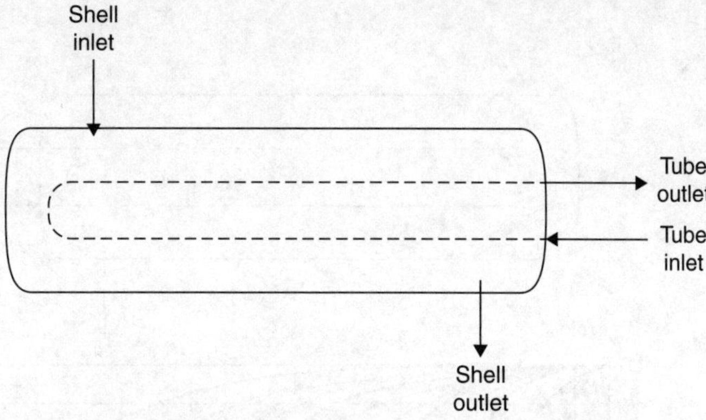

FIGURE 8.17   One-shell pass and two-tube passes.

The heat transfer rate can be computed for shell and tube heat exchangers using Eq. (8.12).

$$Q = U \times A \times \Delta T_m$$

where F is a correction factor dependent on the number of shell and tube passes, and $\Delta T_m = F \times LMTD$, where $LMTD = \dfrac{\Delta T_2 - \Delta T_1}{\ln\left(\dfrac{\Delta T_2}{\Delta T_2}\right)}$.

Typical correction factors, "F," are shown in Figs. 8.19 and 8.20.

Example 8.5   A two-shell pass and a four-tube pass heat exchanger is used to heat oil from 30 to 50°C using hot water, which flows through 25-mm-diameter tubes. The hot water enters the heat exchanger at 70°C and leaves at 40°C. The total length of the

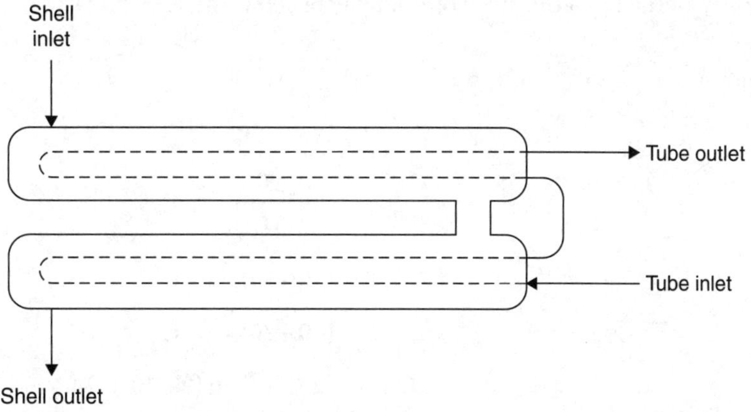

FIGURE 8.18   Two-shell passes and four-tube passes.

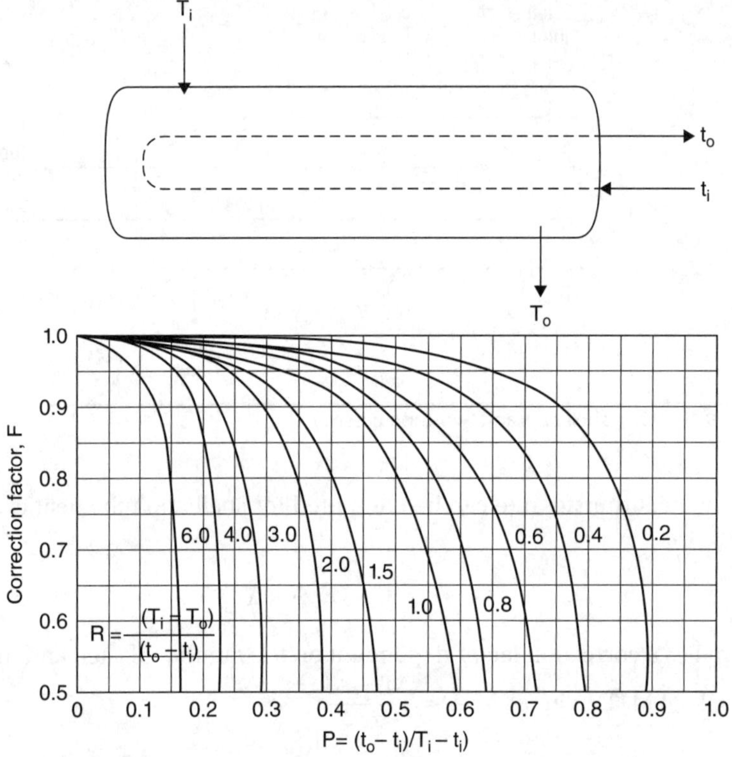

Figure 8.19   Correction factor F for one-shell and multiple-tube passes (multiples of 2).

tubes in the heat exchanger is 40 m. The convection heat transfer coefficient is 50 W/m² · K on the oil (shell) side and 200 W/m² · K on the water (tube) side. Determine the rate of heat transfer in the heat exchanger if the tube thickness is 0.5 mm and the tube material has a thermal conductivity of 400 W/m · K.

Solution

$$\text{LMTD} = \frac{\Delta T_2 - \Delta T_1}{\ln\left(\dfrac{\Delta T_2}{\Delta T_2}\right)}$$

$$\Delta T_2 = (70 - 50) = 20°C$$

$$\Delta T_1 = (40 - 30) = 10°C$$

$$\text{LMTD} = (20 - 10)/\ln(20/10) = 14.4°C$$

$$\text{LMTD}_{\text{shell and tube}} = F \times \text{LMTD}_{\text{counterflow}}$$

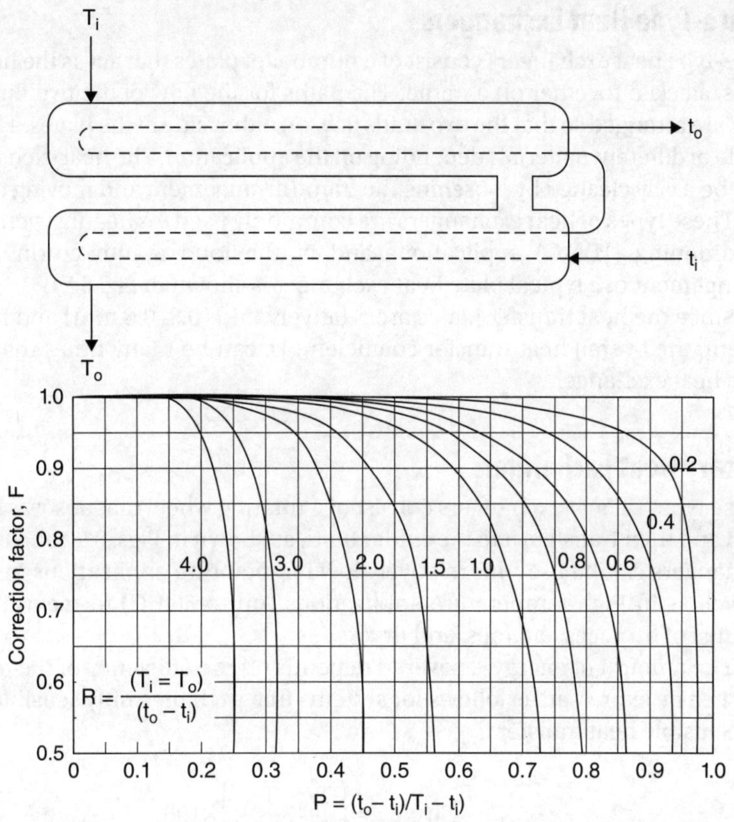

Figure 8.20 Correction factor F for two-shell and multiple-tube passes (multiples of 4).

From Fig. 8.20:

$$P = (t_o - t_i)/(T_i - t_i) = 30/40 = 0.75$$

$$R = (T_i - T_o)/(t_o - t_i) = 20/30 = 0.67$$

$$F = 0.85$$

Tube surface area $= \pi \times 0.025 \times 40 = 3.14 \text{ m}^2$

$$U = \left(\cfrac{1}{\dfrac{1}{h_A} + \dfrac{x_1}{k_1} + \dfrac{1}{h_B}} \right) \quad [\text{from Eq. (8.6)}]$$

$$1/U = (1/50) + (0.0005/400) + (1/200)$$

$$U = 40 \text{ W/m}^2 \cdot \text{K}$$

$$Q = U \times A \times \text{LMTD} \times F$$

$$Q = 40 \times 3.14 \times 14.4 \times 0.85 = 1537 \text{ W} \quad \blacktriangle$$

8.6   Plate-Type Heat Exchangers

Plate-type heat exchangers consist of a number of plates that act as the heat transfer surfaces, stacked together on a frame. The paths for the flow of the two fluids transferring heat are arranged so that the two fluids flow on either side of the plates. The plates can be made of different materials depending on the application. The heat exchanger plates can also be easily cleaned by loosening the frame arrangement and moving the plates.

These types of heat exchangers are commonly used for heating, ventilating, and air conditioning (HVAC) applications and in the food manufacturing industry. The arrangement of a typical plate heat exchanger is shown in Fig. 8.21.

Since the heat transfer plates are relatively thin (0.3–0.6 mm) and fluid flow is turbulent, the overall heat transfer coefficient, U, can be many times that of a shell and tube heat exchanger.

8.7   Rotary Heat Exchangers

These types of heat exchangers consist of a rotating wheel that absorbs heat from a hot fluid and then transfers it to the colder fluid, as shown in Figs. 8.22 and 8.23. These heat exchangers can be used for relatively low temperature applications in HVAC systems as well as in high-temperature applications (up to 500°C) to recover heat from the exhaust of furnaces, engines, and dryers.

The amount of energy recovered depends on the efficiency of the recovery system and can be expressed as follows for systems that exchange only sensible heat.

Sensible heat transfer:

$$\eta_{\text{Sensible}} = \left(\frac{T_{SA} - T_{OA}}{T_{DA} - T_{OA}} \right) \times 100 \tag{8.16}$$

where η = efficiency (%), T = dry-bulb temperature (°C), OA = outdoor air, SA = supply air, and DA = discharge air (exhaust from the industrial process).

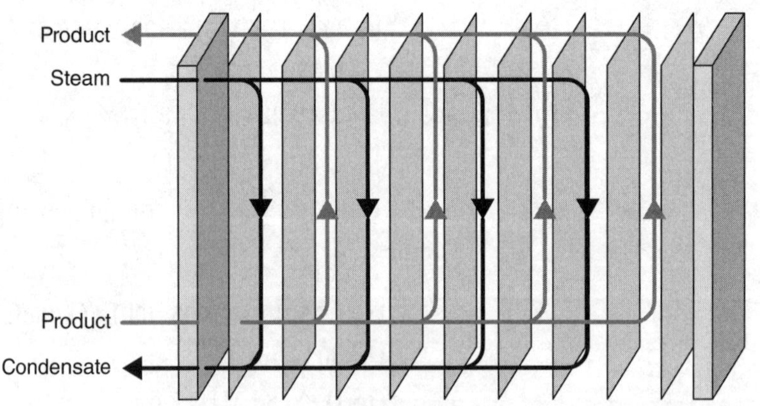

Figure 8.21   Arrangement of a plate heat exchanger (courtesy of Spirax Sarco).

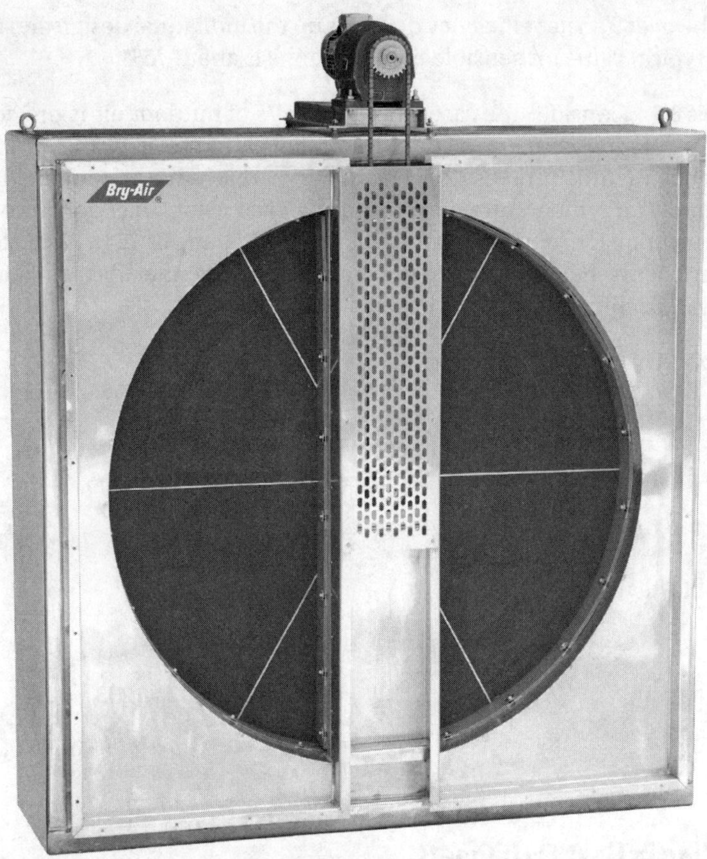

Figure 8.22   Rotary heat exchanger (courtesy of Bry-Air).

The amount of energy recovered can be expressed as follows:

$$Q_{sensible} = \rho \times v \times C_p \times (T_{SA} - T_{OA}) \tag{8.17}$$

where Q = energy recovered (kW), ρ = density of air (kg/m^3), v = airflow rate (m^3/s), and C_p = specific heat capacity of air (kJ/kg · K).

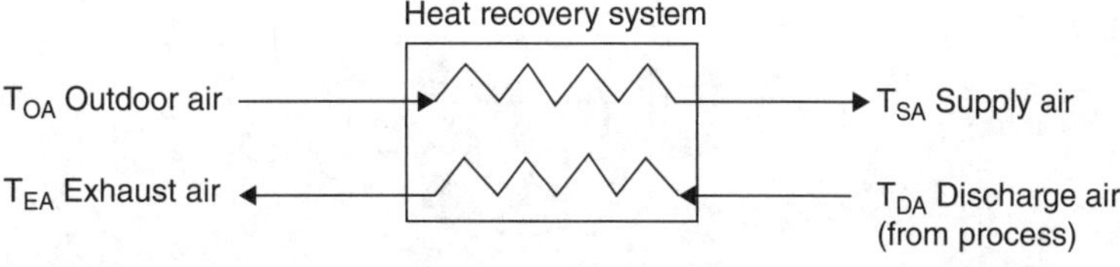

Figure 8.23   Heat exchange in a rotary heat exchanger.

The heat transfer efficiency depends on the individual design and the flow velocity, but a typical value for sensible heat exchange is about 75%.

Example 8.6   Consider the case where 1.2 m³/s of outdoor air is provided to an industrial dryer to make up for an equal amount of air removed by the exhaust system. If the outdoor air temperature is 32°C and the dryer discharge air temperature is 300°C, find the supply air temperature that can be achieved using an energy recovery system that can only transfer sensible heat and has an efficiency of 75%. Also, calculate the total amount of pre-heating done to the outdoor air. Take the density of air to be 0.9 kg/m³ and the specific heat capacity of air as 1002 J/kg·K.

Solution   Using Eq. (8.16):

$$T_{SA} = T_{OA} + (T_{DA} - T_{OA}) \times \eta_{Sensible}$$

$$= 32 + (300 - 32) \times 0.75$$

$$= 233°C$$

Sensible heat added:

$$Q_{sensible} = \rho \times v \times C_p \times (T_{SA} - T_{OA})$$

$$= 0.9 \times 1.2 \times 1002 \times (233 - 32)$$

$$= 217.5 \text{ kW} \quad \blacktriangle$$

8.8   Fouling in Heat Exchangers

Heat exchanger surfaces are often subjected to fouling and scaling due to the deposition of fluid impurities, rust formation, or chemical reactions between the fluid and wall material. Formation of a fouling or scaling on the heat transfer surface creates an additional layer on the surface, resulting in a greater resistance to heat transfer.

Fouling depends on factors such as the type of fluids used, materials used for construction of the heat exchanger, and length of time the heat exchanger has been in operation. Often, heat exchangers need to be cleaned periodically to minimize the resistance to heat transfer resulting from fouling.

Therefore, the overall heat transfer coefficient value (U) given in Eq. (8.6) should be adjusted to include the effect of fouling, as shown in Eq. (8.18), to ensure that the heat exchanger selected for a particular application is able to provide the required capacity.

$$U = \left(\cfrac{1}{\cfrac{1}{h_A} + \cfrac{x_1}{k_1} + \cfrac{x_2}{k_2} + \cfrac{1}{h_B} + R_f} \right) \tag{8.18}$$

where R_f is the fouling factor (zero for clean surfaces).

Fluid	Fouling factor R_f ($m^2 \cdot K/W$)
Treated water from cooling tower	0.0003
River water	0.0004
Sea water	0.0002
Fuel oil	0.0009
Steam	0.0001
Refrigerating liquids	0.0002
Treated boiler feedwater	0.0002
Boiler blowdown	0.00035
Hydraulic and heat transfer oil	0.00018
Cooling fluids	0.00018

TABLE 8.2   Typical Fouling Factors

Some typical fouling factors used when selecting and designing heat exchangers are provided in Table 8.2.

Example 8.7   Calculate the new heat transfer rate for Example 8.5, taking the fouling factor to be 0.0002 $m^2 \cdot K/W$.

Solution

$$U = \left(\frac{1}{\dfrac{1}{h_A} + \dfrac{x_1}{k_1} + \dfrac{1}{h_B} + R_f} \right) \quad [\text{from Eq. (8.18)}]$$

$$1/U = (1/50) + (0.0005/400) + (1/200) + 0.0004$$

$$U = 39.4 \; W/m^2 \cdot K$$

$$Q = U \times A \times LMTD \times F$$

$$Q = 39.4 \times 3.14 \times 14.4 \times 0.85 = 1514 \; W \quad \blacktriangle$$

8.9   Overall Heat Transfer Coefficient Values

Computation of the overall heat transfer coefficient for a particular heat exchanger requires the thermal conductivity value of the heat exchanger material, its thickness, and the heat transfer coefficients at the inner and outer surfaces where heat transfer takes place. Although the first two parameters can be obtained from the design specifications of a heat exchanger, computation of the surface heat transfer coefficient values requires in-depth heat transfer analysis as they depend on properties of the fluid, flow velocity, and surface conditions.

Type of heat exchanger	Fluids	Overall heat transfer coefficient, U (W/m²·k)
Shell and tube	Water to water	800–1500
	Light oil to light oil	100–400
	Heavy oil to heavy oil	50–300
	Solvent to solvent	100–300
Plate heat exchanger	Liquid to liquid	5000–7500
	Steam to water	3500–4500
Double pipe	Liquid to liquid	150–1200
	Liquid inside, steam outside	300–1200

TABLE 8.3   Typical Heat Transfer Coefficient Values

Therefore, to facilitate ease of computation, typical values for the overall heat transfer coefficient are provided in many reference sources. Some typical values are listed in Table 8.3.

As can be seen from Table 8.3, the U values depend on the respective fluids and the type of heat exchanger. In general, for the three types of heat exchangers compared in the table, double-pipe heat exchangers have the lowest U values, while plate-type heat exchangers have the highest U values.

CHAPTER 9

Combined Heat and Power Systems

9.1 Introduction

Cogeneration usually refers to the simultaneous generation of multiple forms of useful energy from a single source of primary energy. The most common types of cogeneration systems are those which generate electricity and heat simultaneously and are called combined heat and power (CHP) systems.

The main benefit of CHP systems is the significantly higher overall efficiency when compared to conventional power generation systems. In conventional power plants, only a fraction of the energy potential in fuel is converted into electrical energy, as illustrated in Fig. 9.1.

The typical operating efficiency of common electricity generation systems using internal combustion (IC) engines, gas turbines, steam turbines, and combined cycle plants ranges from about 30% to 55%.

9.2 Internal Combustion Engine

An IC engine coupled to an alternator is an arrangement widely used for small- to medium-scale power generation. In such systems, only about 30% to 40% of the energy content of the fuel is converted into useful electrical energy as most of the input energy is wasted as heat from the exhaust and jacket cooling.

IC engines commonly operate either on the Otto cycle (spark ignition) or on the Diesel cycle (compression ignition). The construction of both types of engines is similar in that each has a piston moving in a cylinder to compress air (or a mixture of air and fuel), which is then ignited, resulting in the gas to expand and move the piston (power stroke). Multiple pistons are connected to a crankshaft and the operating cycle of each piston is staggered to achieve a power stroke at different times, resulting in the rotation of the crankshaft. The engine crankshaft is coupled to an electrical generator.

The main difference between the two cycles is that in the Otto cycle the air and fuel mixture is compressed and ignited using a spark plug, while in the Diesel cycle, fuel is injected into the cylinder once the air is compressed and the temperature has reached the ignition temperature, thereby igniting the mixture.

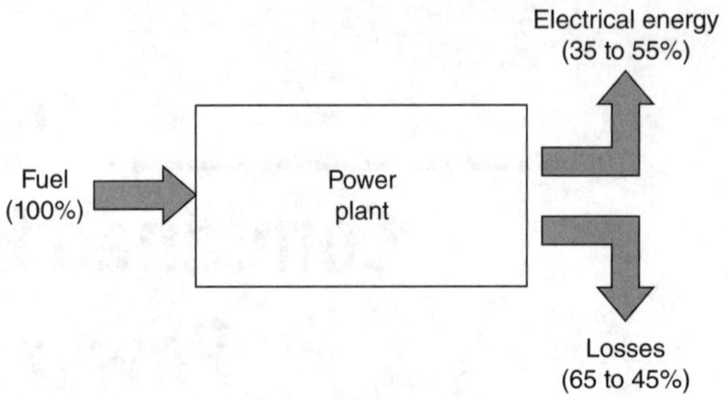

Figure 9.1   Efficiency of conventional power plants.

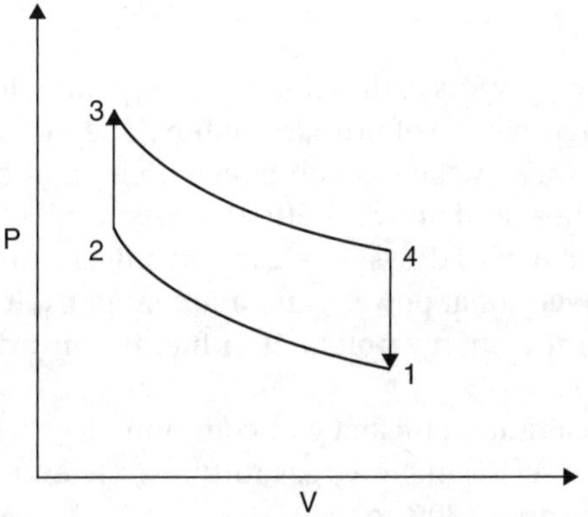

Figure 9.2   P–V diagram for the Otto cycle.

The pressure versus volume diagram for the Otto cycle is shown in Fig. 9.2. The cycle consists of the following four processes:

- 1 to 2: isentropic compression
- 2 to 3: reversible constant volume heating
- 3 to 4: isentropic expansion
- 4 to 1: reversible constant volume cooling

Similarly, the pressure versus volume diagram for the Diesel cycle is shown in Fig. 9.3. The cycle consists of the following four processes:

- 1 to 2: isentropic compression
- 2 to 3: reversible constant pressure heating
- 3 to 4: isentropic expansion
- 4 to 1: reversible constant volume cooling

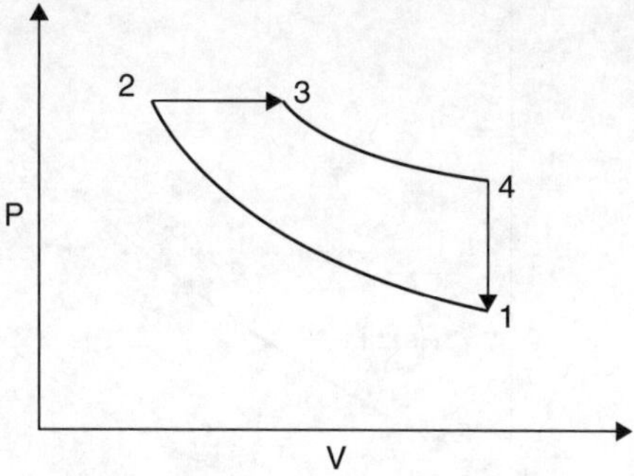

FIGURE 9.3 P–V diagram for the Diesel cycle.

9.3 Gas Turbine

Another type of power generation plant used for medium-size generation consists of a gas turbine coupled to an alternator. In an open cycle gas turbine (Fig. 9.4), air enters the compressor, which is powered by the gas turbine. After compression, the air enters the combustion chamber where fuel is introduced. After combustion, the high-temperature and high-pressure gas is expanded in the turbine. Typical generation efficiency is about 30%.

The gas turbine operates on the thermodynamic Brayton cycle. The temperature versus entropy (a thermodynamic property), called a T–S diagram, for the open cycle gas turbine is shown in Fig. 9.5.

The ideal cycle consists of the following four processes:

- 1 to 2: isentropic compression
- 2 to 3: isobaric heating
- 3 to 4: isentropic expansion
- 4 to 1: isobaric cooling (applicable for closed cycle)

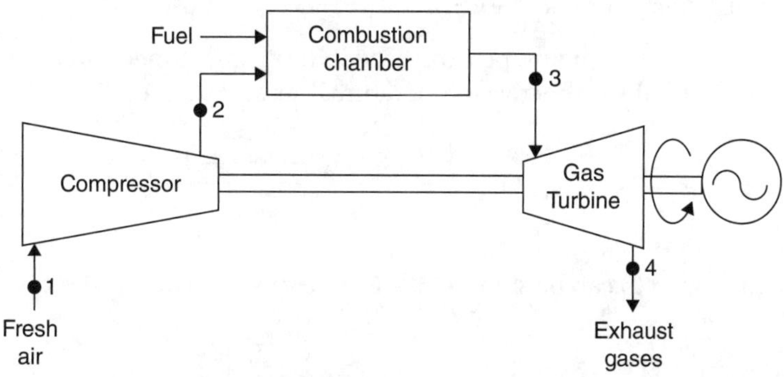

FIGURE 9.4 Arrangement of an open cycle gas turbine.

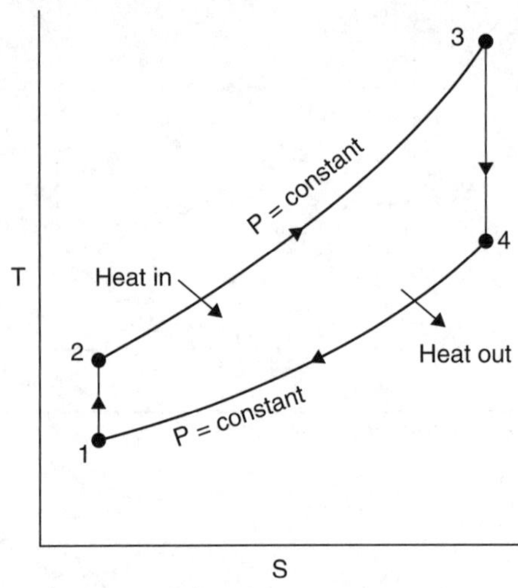

Figure 9.5   T–S diagram for an ideal gas turbine cycle.

The overall performance of the cycle can be analyzed using the following relationships:

$$\text{Heat input, } Q_{in} = Q_{23} = m \times (h_3 - h_2)$$

$$\text{Compressor work done, } W_{comp} = m \times (h_2 - h_1)$$

$$\text{Turbine work done, } W_{turbine} = m \times (h_3 - h_4)$$

$$\text{Net work done, } W_{net} = W_{turbine} - W_{comp}$$

$$\text{Thermal efficiency, } \eta_{th} = W_{net}/Q_{in}$$

$$\eta_{th} = [(h_3 - h_4) - (h_2 - h_1)]/(h_3 - h_2) \tag{9.1}$$

where h refers to the specific enthalpy of air at a particular operating condition and m is the mass of air. Enthalpy is a thermodynamic property and is used to represent the energy content of a unit mass of a substance. It cannot be measured directly but values are published in reference tables.

The efficiency can also be expressed in terms of temperature, T, and specific heat capacity, C_p (at the respective temperatures), as shown in Eq. (9.2):

$$\eta_{th} = \left[\frac{Cp(T_3 - T_4) - Cp(T_2 - T_1)}{Cp(T_3 - T_2)} \right] \tag{9.2}$$

Equation (9.2) can be simplified if C_p is assumed to be constant:

$$\eta_{th} = 1 - \left(\frac{T_4 - T_1}{T_3 - T_2} \right) \tag{9.3}$$

Usually, the inlet temperature to the compressor (T_1) and inlet temperature to the turbine (T_3) are known. Since the compressor and turbine pressure ratios are also known for a particular system, the temperature at the exit of the compressor and turbine can be computed using the following relationships:

$$\frac{T_2}{T_1} = \left(\frac{P_2}{P_1}\right)^{\left(1-\frac{1}{\gamma}\right)} \text{ and } \frac{T_3}{T_4} = \left(\frac{P_3}{P_4}\right)^{\left(1-\frac{1}{\gamma}\right)} \tag{9.4}$$

The thermal efficiency of the cycle increases with increase in the pressure ratio. However, increase in pressure ratio beyond a point leads to lower net power output (due to increased power required at the compressor). When the net power output reduces, for a particular capacity of a power generation plant, the mass flow rate has to be increased, leading to a bigger size plant. Therefore, pressure ratios are generally maintained between 11 and 16 to optimize the efficiency and size of gas turbines.

Example 9.1   The air temperature at the compressor inlet is 32°C and the pressure ratio of the gas turbine is 7. If the gas temperature at the inlet to the turbine is 1007°C, compute the operating thermal efficiency of the gas turbine. Take the specific heat capacity of air and exhaust gas to be 1.005 and 1.15 KJ/kg·K, respectively. The value of γ can be taken as 1.4 for both the compression and expansion processes. If the mass flow rate of air is 2.5 kg/s, compute the net power output.

Solution

$$T_1 = 273 + 32 = 305 \text{ K}$$

$$\text{From Eq. (9.4), } T_2 = T_1\left(\frac{P_2}{P_1}\right)^{\left(1-\frac{1}{\gamma}\right)} = 305 \times (7)^{0.286} = 532 \text{ K}$$

$$\text{At the turbine inlet, } T_3 = 1007 + 273 = 1300 \text{ K}$$

$$\text{From Eq. (9.4), } T_4 = 1300/(7)^{0.286} = 745 \text{ K}$$

From Eq. (9.2):

$$\eta_{th} = \left[\frac{Cp(T_3 - T_4) - Cp(T_2 - T_1)}{Cp(T_3 - T_2)}\right]$$

$$= \left[\frac{1.15 \times (1300 - 745) - 1.005 \times (532 - 305)}{1.15 \times (1300 - 532)}\right] = 0.46 \,(46\%)$$

$$\eta_{th} = \text{net power output/input power}$$

$$\text{Since heat input, } Q_{in} = Q_{23} = m \times (h_3 - h_2)$$

$$\text{Power input} = \dot{m} \times (h_3 - h_2) = \dot{m} \times C_p \times (T_3 - T_2)$$

where $\dot{m}$ is the mass flow rate.

$$\text{Input power} = \dot{m} \times C_p \times (T_3 - T_2) = 2.5 \times 1.15 \times (1300 - 532) = 2208 \text{ kW}$$

$$\text{Net power output} = 0.46 \times 2208 = 1015 \text{ kW} \quad \blacktriangle$$

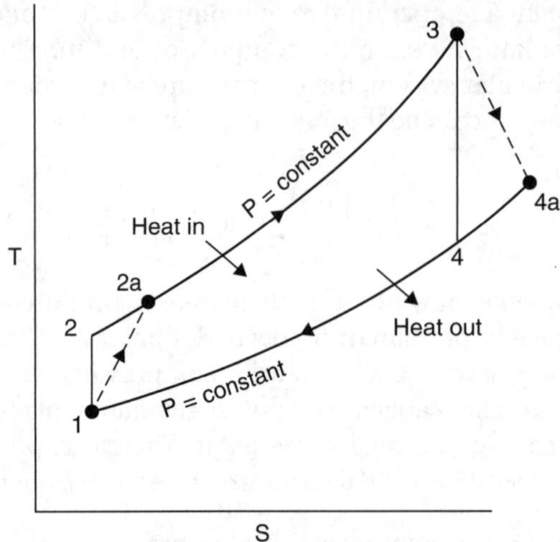

FIGURE 9.6   T–S diagram for a gas turbine cycle with isentropic efficiencies.

In the above example, it was assumed that the compression and expansion processes are isentropic (constant entropy). However, in actual operation, these processes are not isentropic, as illustrated in Fig. 9.6.

The isentropic efficiency for compression and expansion processes can be expressed as follows (taking the specific heat capacity to be constant):

$$\eta_{i,comp} = (T_2 - T_1)/(T_{2a} - T_1) \tag{9.5}$$

$$\eta_{i,turb} = (T_3 - T_{4a})/(T_3 - T_4) \tag{9.6}$$

Example 9.2   Taking the isentropic efficiency of the compressor and turbine to be 80% and 83% in Example 9.1, compute the overall thermal efficiency and the net power output.

Solution   From Eq. (9.5):

$$\eta_{i,comp} = 0.8 = (532 - 305)/(T_{2a} - 305)$$

Therefore, $T_{2a} = 589$ K

Similarly, $\eta_{i,turb} = 0.83 = (1300 - T_{4a})/(1300 - 745)$

Therefore, $T_{4a} = 839$ K

New thermal efficiency, $\eta_{th} = \left[\dfrac{Cp(T_3 - T_{4a}) - Cp(T_{2a} - T_1)}{Cp(T_3 - T_{2a})} \right]$

$$= \left[\frac{1.15 \times (1300 - 839) - 1.005 \times (589 - 305)}{1.15 \times (1300 - 589)} \right]$$

$$= 0.3 \text{ (or 30\%)} \quad \blacktriangle$$

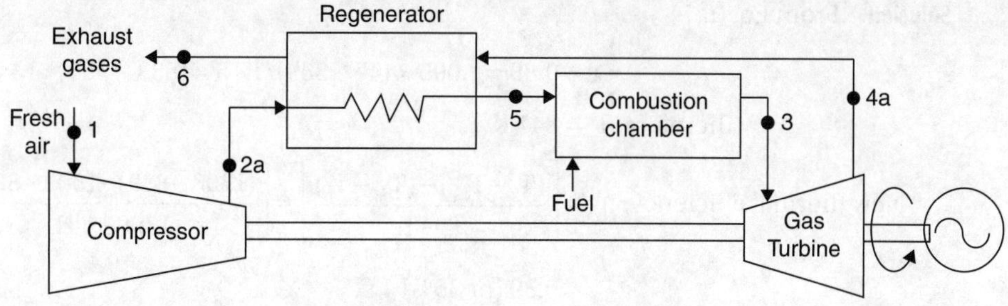

FIGURE 9.7   Gas turbine with regeneration.

The above example shows that when the isentropic efficiency of the compressor and turbine is considered, the overall thermal efficiency drops. Hence, to achieve a relatively better efficiency value, other enhancements such as regeneration are used.

Figure 9.7 shows a turbine with regeneration where some of the waste heat at the exit of the turbine is used to heat air after compression by means of a heat exchanger, so that less heat needs to be added during the combustion process.

The T–S diagram for the cycle with regeneration is shown in Fig. 9.8.

The effectiveness of the regenerator can be expressed as:

$$\varepsilon = (h_5 - h_{2a})/(h_{4a} - h_{2a}) = C_{pa}(T_5 - T_{2a})/C_{pg}(T_{4a} - T_{2a}) \tag{9.7}$$

where C_{pa} and C_{pg} are the specific heat capacities of air and exhaust gas, respectively.

Example 9.3   Compute the achievable thermal efficiency and the net power output for the gas turbine system in Example 9.2 with a regeneration system having an effectiveness of 0.88. The specific heat capacity of the exhaust gas can be taken as 1.15 kJ/kg·K

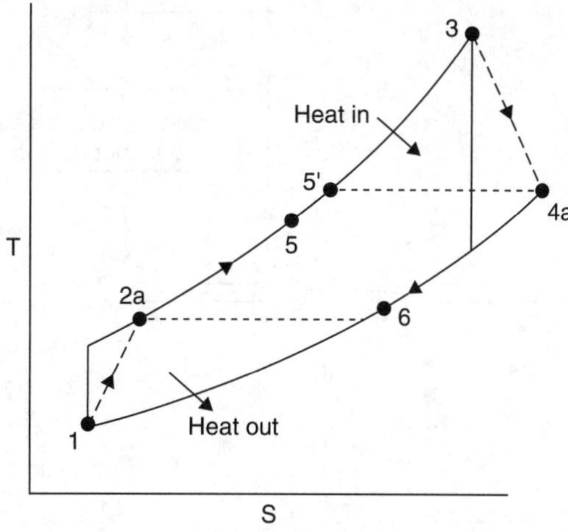

FIGURE 9.8   T–S diagram for a gas turbine with regeneration.

Solution   From Eq. (9.7)

$$\varepsilon = 0.88 = 1.005 \times (T_5 - 589)/1.15 \times (839 - 589)$$

Therefore, $T_5 = 841$ K

$$\text{New thermal efficiency, } \eta_{th} = \left[\frac{(T_3 - T_{4a}) - (T_{2a} - T_1)}{T_3 - T_5}\right] = \left[\frac{(1300 - 839) - (589 - 305)}{1300 - 841}\right]$$

$$= 0.39 \text{ (or 39\%)}$$

(assuming C_p is constant)

If the two values of C_p are used (more accurate estimation):

$$\eta_{th} = \left[\frac{Cp_g (T_3 - T_{4a}) - Cp_a (T_{2a} - T_1)}{Cp_g (T_3 - T_5)}\right] = \left[\frac{1.15 \times (1300 - 839) - 1.005 \times (589 - 305)}{1.15 \times (1300 - 841)}\right]$$

$$= 0.46 \text{ (or 46\%)} \quad \blacktriangle$$

The efficiency of gas turbine systems can also be improved by using multistage compression with intercooling and multistage expansion with reheating. The arrangement of such a system is shown in Fig. 9.9, while the respective T–S diagram is shown in Fig. 9.10.

Intercooling between stages of compression helps to reduce the work input, while reheating between stages of expansion helps to increase the work output, both of which help to improve the overall thermal efficiency.

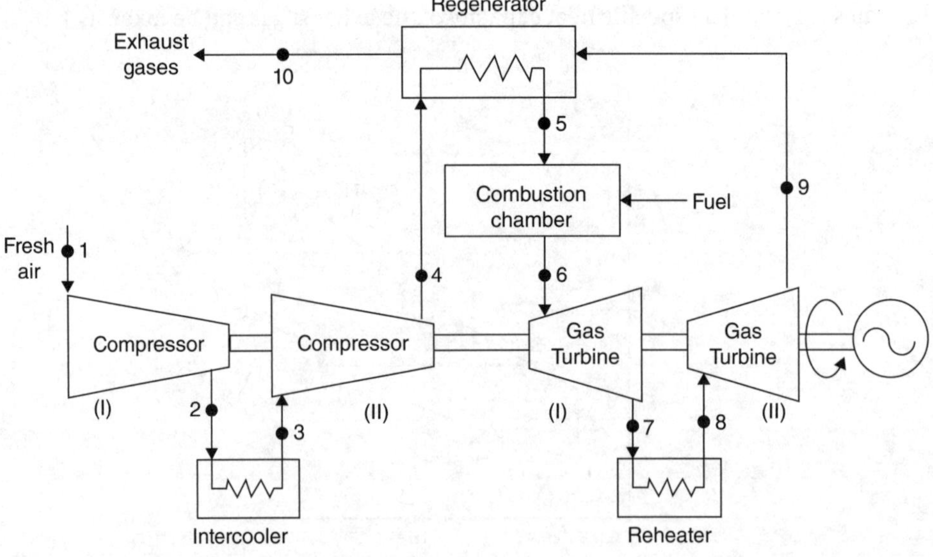

FIGURE 9.9   Gas turbine with intercooling and reheating.

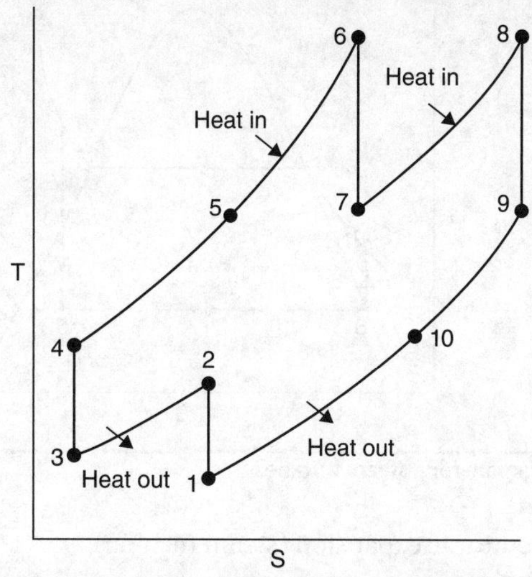

FIGURE 9.10 T–S diagram for a gas turbine with intercooling and reheating.

9.4 Steam Turbine

Another type of power generation cycle uses steam turbines, as shown in Fig. 9.11. High-pressure steam is generated in a boiler using a variety of different fuels such as oil, gas, coal, or biomass and expanded in a steam turbine, which is coupled to an alternator. The low-pressure steam exiting the turbine is condensed and fed back to the boiler. The typical operating efficiency of this cycle is about 45%.

The steam turbine cycle operates on the thermodynamic Rankine cycle. The temperature versus entropy (T–S) diagram for a steam turbine system is shown in Fig. 9.12.

The cycle consists of the following four processes:

- 6 to 1: isentropic compression (pump)

- 1 to 3: isobaric heating (boiler)

- 3 to 4: superheating of steam

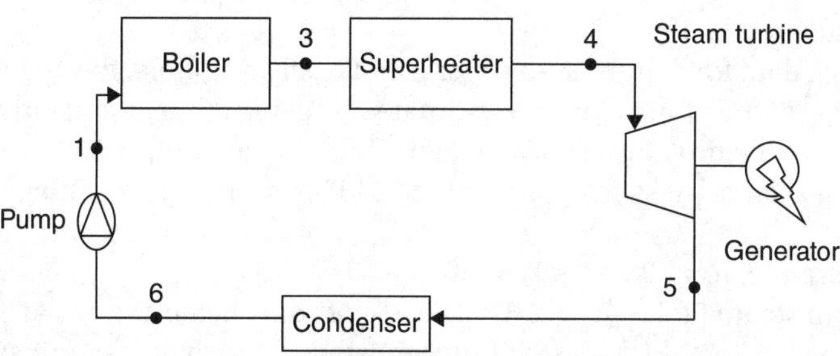

FIGURE 9.11 Arrangement of a steam turbine power generation plant.

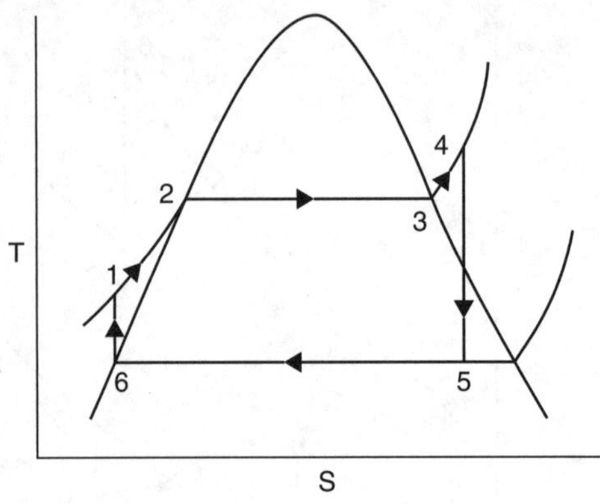

Figure 9.12 T–S diagram for a steam turbine.

- 4 to 5: isentropic expansion (steam turbine)
- 5 to 6: isobaric cooling (condenser)

The work and heat inputs and work output can be expressed in terms of specific enthalpies at the different points and the mass flow rate.

$$\text{Output from the turbine} = m \times (h_4 - h_5)$$

$$\text{Work input to pump} = m \times (h_1 - h_6)$$

$$\text{Heat input} = m \times (h_4 - h_1)$$

where m is the mass and h is the specific enthalpy.

The thermal efficiency of the system, η = net work output/heat input:

$$\eta = [m\,(h_4 - h_5) - m\,(h_1 - h_6)]/m\,(h_4 - h_1)$$

$$\eta = [(h_4 - h_5) - (h_1 - h_6)]/(h_4 - h_1) \tag{9.8}$$

Example 9.4   The operation of a steam power plant is shown in Fig. 9.13. Compute the thermal efficiency of the plant.

Solution   Specific enthalpy at inlet to pump, $h_6 = h_f$ at 0.1 bar (10 kPa) = 192 kJ/kg·K (from steam tables)

Since data for compressed water is not easily available, the specific enthalpy at the discharge of the pump h_1 can be estimated by using the relationship $h_1 \approx h6 + v\,(P_1 - P_6)$, where v = specific volume of water at the inlet to the pump = 0.001 m³/kg at 0.1 bar, P_1 = pump discharge pressure = 30 bar = 3000 kPa, and P_6 = pump inlet pressure = 0.1 bar = 10 kPa.

Therefore, $h_1 = 192 + 0.001 \times (3000 - 10) = 195$ kJ/kg.

From steam tables, $h_4 = 3231$ kJ/kg (h_g of superheated steam at 30 bar and 400°C).

Also, $S_4 = 6.921$ kJ/kg·K (S_g of superheated steam at 30 bar and 400°C).

Since the expansion process is isentropic, $S_5 = S_4 = 6.921$ kJ/kg·K.

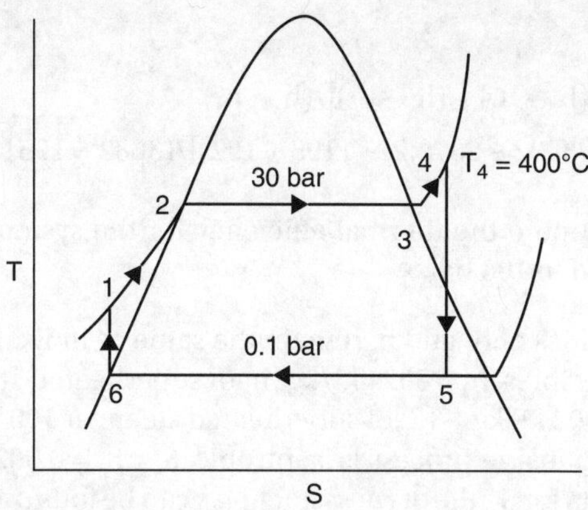

FIGURE 9.13   T–S diagram for Example 9.4.

Using the data for S_5, the dryness fraction x can be found at 0.1 bar since $S_5 = S_f + x(S_{fg})$

$$\text{Therefore, } 6.921 = 0.649 + x(7.5)$$

$$\text{Hence, } x = 0.836$$

Specific enthalpy at the exit of the turbine, $h_5 = h_f + x(h_{fg})$ (at 0.1 bar)

$$\text{Therefore, } h_5 = 192 + 0.836 \times 2392 = 2191.7 \text{ kJ/kg·K}$$

Using Eq. (9.8):

$$\eta = [(h_4 - h_5) - (h_1 - h_6)]/(h_4 - h_1)$$
$$= [(3231 - 2191.7) - (195 - 192)]/(3231 - 195) = 0.34 \text{ (or 34\%)} \quad \blacktriangle$$

The thermal efficiency of the system in Example 9.4 is relatively low. The efficiency of steam turbine power plants can be improved by lowering the condensing pressure (which is limited by the temperature of the condenser cooling medium), increasing superheat (Example 9.5), and increasing the operating system pressure (Example 9.6).

Example 9.5   Compute the thermal efficiency for the system in Example 9.5 with the superheat increased from 400 to 600°C.

Solution   The values of h_1 and h_6 remain the same as in Example 9.4.
From steam tables, $h_4 = 3682$ kJ/kg (h_g of superheated steam at 30 bar and 600°C).
Also, $S_4 = 7.507$ kJ/kg·K (S_g of superheated steam at 30 bar and 600°C).
Since the expansion process is isentropic, $S_5 = S_4 = 7.506$ kJ/kg·K.
Using the data for S_5, the dryness fraction x can be found at 0.1 bar since $S_5 = S_f + x(S_{fg})$.

Therefore, $7.507 = 0.649 + x(7.5)$

Hence, $x = 0.914$

Specific enthalpy at the exit of the turbine, $h_5 = h_f + x \cdot (h_{fg})$ (at 0.1 bar)

Therefore, $h_5 = 192 + 0.914 \times 2392 = 2379.2$ kJ/kg·K

Using Eq. (9.8):

$$\eta = [(h_4 - h_5) - (h_1 - h_6)]/(h_4 - h_1)$$
$$= [(3682 - 2379.2) - (195 - 192)]/(3682 - 195) = 0.37 \text{ (or 37\%)} \quad \blacktriangle$$

Example 9.6   Compute the thermal efficiency for the system in Example 9.5 when the pressure is raised to 100 bar.

Solution   The values of h_1 and h_6 remain the same as in Example 9.5.
From steam tables, $h_4 = 3624$ kJ/kg (h_g of superheated steam at 100 bar and 600°C).
Also, $S_4 = 6.902$ kJ/kg·K (S_g of superheated steam at 100 bar and 600°C).
Since the expansion process is isentropic, $S_5 = S_4 = 6.902$ kJ/kg·K.
Using the data for S_5, the dryness fraction x can be found at 0.1 bar since $S_5 = S_f + x(S_{fg})$.

$$\text{Therefore, } 6.902 = 0.649 + x(7.5)$$

$$\text{Hence, } x = 0.834$$

Specific enthalpy at the exit of the turbine, $h_5 = h_f + x(h_{fg})$ (at 0.1 bar)

$$\text{Therefore, } h_5 = 192 + 0.834 \times 2392 = 2186.3 \text{ kJ/kg·K}$$

Using Eq. (9.8):

$$\eta = [(h_4 - h_5) - (h_1 - h_6)]/(h_4 - h_1)$$
$$= [(3624 - 2186.3) - (195 - 192)]/(3624 - 195) = 0.42 \text{ (or 42\%)} \quad \blacktriangle$$

As seen from Examples 9.5 and 9.6, the thermal efficiency improves when the superheat and operating pressure are increased. However, it should be noted that there are limitations to increasing the operating temperature and pressures due to metallurgical considerations. Present state-of-the-art plants operating under ultrasupercritical conditions operate at steam pressures between 250 and 290 bar and temperatures in the range of 600 to 620°C. The next generation of steam plants using nickel alloy for tubing, piping, and turbine forgings are expected to operate at temperatures of 700 to 720°C and achieve an overall net plant efficiency of 50%.

Meanwhile, another means of increasing the thermal efficiency using the present technology is by expanding in two stages with reheating in between, as illustrated in Fig. 9.14, where steam at the exit of the high-pressure turbine is reheated before entering the low-pressure turbine. The T-S diagram for a steam cycle with reheat is shown in Fig. 9.15.

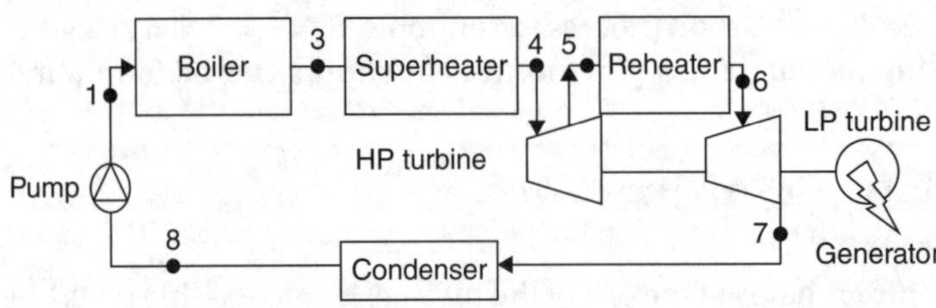

FIGURE 9.14   Steam power plant with reheat.

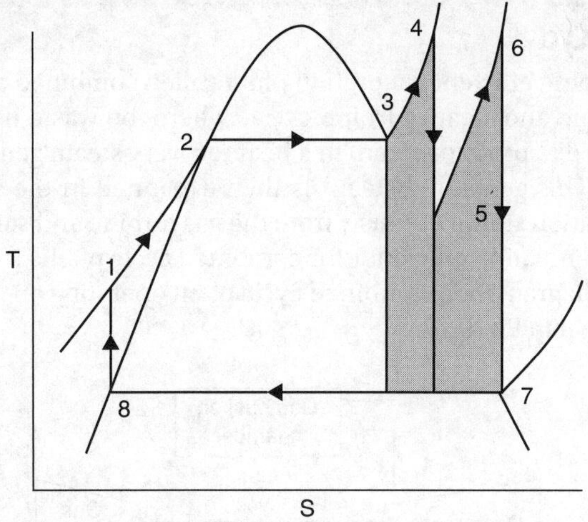

FIGURE 9.15 T–S diagram for a steam cycle with reheat.

Example 9.7 Consider the steam power generation system in Example 9.6, which is operating at 100 bar and 600°C and condensing at 0.1 bar. If the steam is expanded in the high-pressure turbine to 40 bar, compute the overall thermal efficiency of the plant.

Solution The values of h_1 and h_4 remain the same as in Example 9.6.

From steam tables, $h_4 = 3624$ kJ/kg (h_g of superheated steam at 100 bar and 600°C)

Also, $S_4 = 6.902$ kJ/kg·K (S_g of superheated steam at 100 bar and 600°C)

Since the expansion process is isentropic, $S_5 = S_4 = 6.902$ kJ/kg·K

From steam tables, the value of h_5 (at 40 bar) can be found by linear interpolation:

$$(S_5 - 6.769)/(6.935 - 6.796) = (h_5 - 3214) - (3330 - 3214)$$

Since $S_5 = 6.902$ kJ/kg·K, h_5 is calculated to be 3325 kJ/kg·K

Also, $h_6 = 3674$ kJ/kg·K and $S_6 = 7.368$ kJ/kg·K (40 bar and 600°C)

Since the expansion process is isentropic, $S_7 = S_6 = 7.368$ kJ/kg·K

Using the data for S_7, the dryness fraction x can be found at 0.1 bar since $S_7 = S_f + x(S_{fg})$

Therefore, $7.368 = 0.649 + x(7.5)$

Hence, $x = 0.896$

Therefore, $h_7 = 192 + 0.896 \times 2392 = 2335.2$ kJ/kg·K

$$\eta = [(h_4 - h_5) + (h_6 - h_7) - (h_1 - h_8)]/[(h_4 - h_1) + (h_6 - h_5)]$$

$$= [(3624 - 3325) + (3674 - 2335.2) - (195 - 192)]/$$
$$[(3624 - 195) + (3674 - 3325)]$$

$$= 0.43\,(43\%) \quad \blacktriangle$$

9.5   Combined Cycle

A relatively more efficient generation plant called combined cycle is a combination of the gas turbine and steam turbine cycles where the waste heat at the exit of the gas turbine is used to produce steam in a heat recovery steam generator (HRSG) as shown in Fig. 9.16. This generated steam is then expanded in the steam turbine, enabling power generation simultaneously from the gas turbine and steam turbine. The achievable power generation efficiency for combined cycle plants is about 55%.

The T–S diagram for a combined cycle plant operating on the Brayton and Rankine cycles is shown in Fig. 9.17.

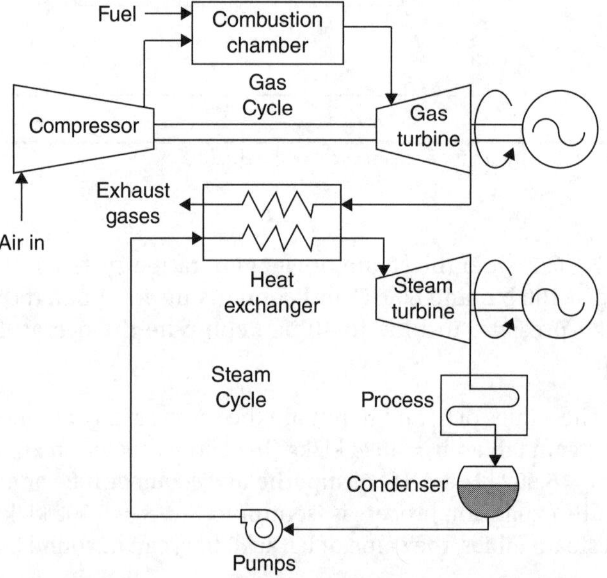

Figure 9.16   Combined cycle plant.

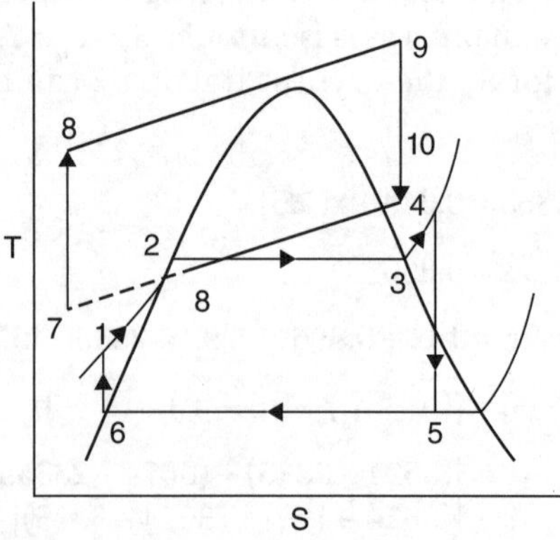

Figure 9.17   T–S diagram for a combined cycle plant.

Example 9.8   Consider the gas turbine cycle in Example 9.2 where the exhaust gas is used to produce steam in a waste HRSG. If the gas turbine exhaust leaves the HRSG at 450°C and if the steam is generated at 7 bar and 200°C and is condensed at 0.1 bar after the steam turbine, compute the mass flow rate of steam and the overall system thermal efficiency. The mass flow rate of air in the gas turbine cycle is 100 kg/s.

Solution   Gas turbine cycle:

$$\text{Input power} = m \times C_p \times (T_9 - T_8) = 100 \times 1.15 \times (1300 - 589) = 82 \text{ MW}$$

$$\text{Net power output} = m \left[C_p \times (T_9 - T_{10}) - C_p \times (T_8 - T_7) \right]$$

$$= 100 \times [1.15 \times (1300 - 839) - 1.005 \times (589 - 305)] = 24.4 \text{ MW}$$

Heat recovery steam generator:

Heat transferred from gas turbine exhaust $= m_{gas} \times C_p \times \Delta h = 100 \times 1.15 \times (839 - 450) = 44.7 \text{ MW}$

The same heat is transferred to steam to increase enthalpy from point 1 to point 4 $= m_{steam} \times (h_4 - h_1)$

where m_{steam} = mass flow rate of steam, h_4 = 2846 kJ/kg·K (superheated steam at 7 bar and 200°C), $h_1 = h_6 +$ specific volume $\times (\Delta P$ of pump), $h_6 = 192$ kJ/kg·K (h_f at 0.1 bar), and $h_1 = 192 + 0.001$ m^3/kg $\times (700$ kPa $- 10$ kPa$) = 192.7$ kJ/kg.

$$\text{Therefore, } 44.7 \text{ MW} = m_{steam} \times (h_4 - h_1) = m_{steam} \times (2846 - 192.7)$$

$$\text{Hence, } m_{steam} = 16.8 \text{ kg/s}$$

Power output from steam turbine:

$$S_4 = 6.888 \text{ kJ/kg·K (superheated steam at 7 bar and 200°C)}$$

Computing dryness fraction x at exit of steam turbine:

$$S_5 = 6.888 = S_f + x(S_{fg}) \text{ (at 0.1 bar)} = 0.649 + x(7.5)$$

$$\text{Therefore, } x = 0.83$$

$$h_5 = h_f + x(h_{fg}) = 192 + 0.83 \times 2392 = 2181.8 \text{ kJ/kg}$$

Net power output from steam turbine = output from steam turbine − power input to pump $= 16.8 \times [(2846 - 2181.8) - (192.7 - 192)] = 11.1 \text{ MW}$

Overall plant efficiency = (output from gas turbine + output from steam turbine)/gas turbine input power

$$= (24 \text{ MW} + 11.1 \text{ MW})/82$$

$$= 0.43 \text{ (or 43\%)}$$

(Note that the overall thermal efficiency has improved from 30% for only the gas turbine to 43% for the combined cycle.)   ▲

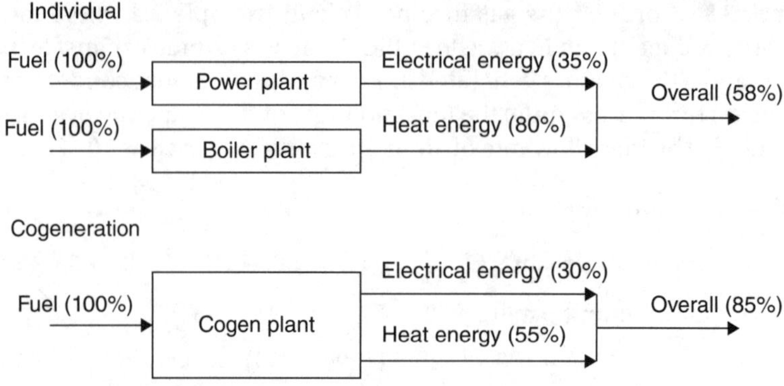

Figure 9.18 Efficiency of CHP systems.

9.6   Need for CHP Systems

The main reason for the relatively low power generation efficiency of the conventional type of plants is that a significant portion of the input energy is converted into heat and is rejected to the environment.

Therefore, if this waste heat can be utilized, the overall system efficiency can be improved and would result in a CHP system where the same fuel source is used to generate electricity as well as heat. Figure 9.18 shows the energy flow for a typical CHP plant, which can lead to an overall efficiency of greater than 85%.

9.7   Types of Cogeneration Systems

CHP systems can be classified as topping cycles or bottoming cycles based on the order of energy use. In topping cycles, electrical energy is first generated followed by thermal energy. The thermal energy produced in a topping cycle is a byproduct of the system and is used for generating steam by means of a heat recovery steam generator or for other heating processes.

In bottoming cycles, fuel is used to first produce high-temperature thermal energy required for industrial processes, such as in steel, ceramic, and petrochemical plants. Thereafter, the thermal energy rejected from the industrial process is used to generate steam and expanded using a turbine to produce electrical energy.

9.7.1   Gas Turbine Topping

A common type of CHP system consists of a gas turbine coupled to an electrical generator and operates on the thermodynamic Brayton cycle, as shown in Fig. 9.19. Air enters a compressor at atmospheric pressure where it is compressed. It then enters a constant-pressure combustion chamber where fuel is injected. The exhaust gases then leave the combustor at a high temperature of around 1200°C and enter a gas turbine where they are expanded to produce mechanical power. The turbine also drives the compressor, and the remaining mechanical power is directed to a generator to produce electrical energy. The hot gases after expansion leave the turbine at a temperature of

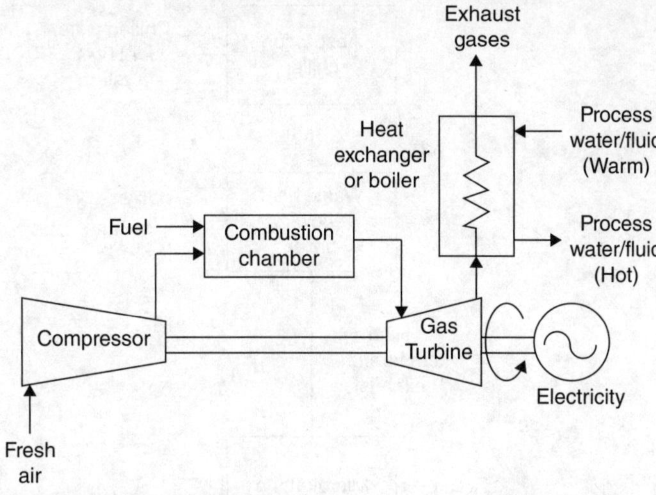

FIGURE 9.19   Arrangement of a gas turbine topping CHP system.

about 450 to 600°C and are then passed through a waste heat boiler to produce steam for heating applications.

Example 9.9   A gas turbine similar to that shown in Fig. 9.19 produces 4.8 MW of electrical power. Natural gas at 1260 m³/h is used by the generator. The water heat from the exhaust is used to generate 10 tons/h of saturated steam at 10 bar. Feedwater is supplied to the heat recovery steam generator at 105°C. Take the calorific value of natural gas to be 10.6 kWh/m³ and specific heat capacity of water to be 4.19 kJ/kg·K. Compute the overall thermal efficiency of the system.

Solution

$$\text{Heat output to steam } Q_{steam} = m_{steam} \times (h_{steam} - h_{feedwater})$$

$$h_{steam} = 2778 \text{ kJ/kg } (h_g \text{ of saturated steam at 10 bar})$$

$$h_{feedwater} = 4.19 \times 105 = 440 \text{ kJ/kg } (h_f \text{ at 105°C})$$

$$Q_{steam} = (10{,}000/3600) \text{ kg/s} \times (2778 - 440) = 6490 \text{ kW}$$

$$\text{Energy input } Q_{fuel} = \text{fuel flow rate} \times \text{calorific value}$$

$$= 1260 \text{ m}^3/\text{h} \times 10.6 \text{ kWh/m}^3 = 13{,}356 \text{ kWh/h} = 13{,}356 \text{ kW}$$

$$\text{Overall thermal efficiency} = (Q_{steam} + \text{electrical output})/Q_{fuel}$$

$$= (6490 + 4800)/13{,}356 = 0.85 \text{ (or 85\%)} \quad \blacktriangle$$

9.7.2   Microturbine

In smaller-size CHP applications, microturbines are used instead of the larger-size gas turbines. Microturbines are essentially a smaller version of combustion turbine generators and are available in output sizes ranging from 25 to 250 kW. They are designed

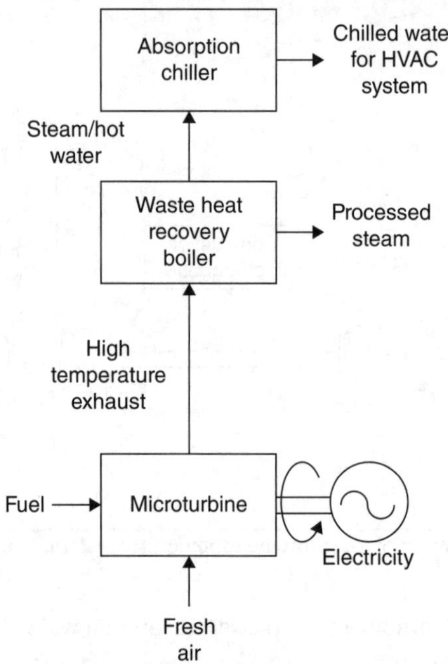

FIGURE 9.20   Arrangement of a microturbine CHP system.

to operate on a variety of fuels such as natural gas, biogas, and propane. Often, recuperators are used to pre-heat the air prior to the combustion chamber using the heat available in the exhaust gases.

Figure 9.20 shows a typical arrangement where air is drawn through a compressor and is pre-heated in a recuperator using hot turbine exhaust gases. The combusted gas is then expanded in a turbine where the thermal energy is converted into mechanical energy. The turbine is coupled to a generator where the mechanical work is converted into electrical energy. The exhaust gas is then used to generate thermal energy for heating applications or cooling using absorption chillers.

9.7.3   Steam Turbine Topping

This cycle operates on the thermodynamic Rankine cycle where high-pressure steam generated in a boiler is expanded in a steam turbine to generate electrical power. The low-pressure steam after expanding is then used for process applications.

In this cycle, depending on the quality of heat required for the process, the steam turbine can operate as a back-pressure or extraction-condensing turbine. In the back-pressure steam turbine, low-pressure steam after the turbine is supplied for thermal process loads shown in Fig. 9.21.

In extraction-condensing steam turbines, part of the steam is extracted from the steam turbine at a pressure required by thermal loads, while the remaining steam is expanded to the lowest pressure possible. The expanded steam is condensed to a low pressure to maximize the power generation by the turbine, as shown in Fig. 9.22.

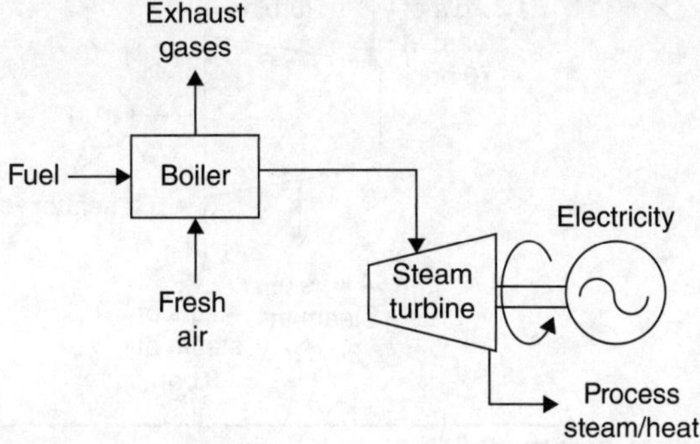

FIGURE 9.21 Back-pressure steam turbine.

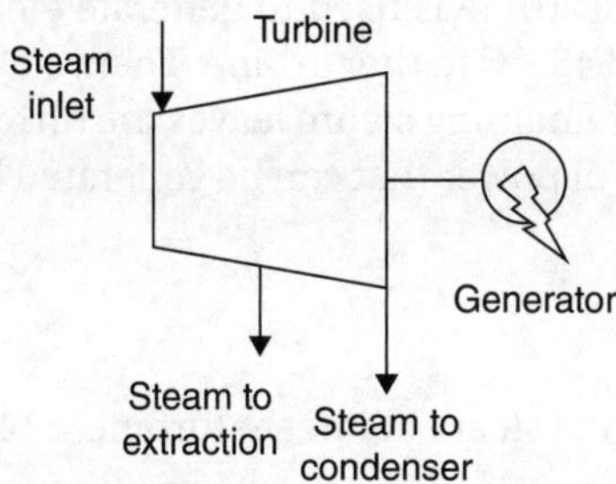

FIGURE 9.22 Extraction-condensing steam turbine.

The power-to-heat ratio for extraction-condensing steam turbines is dependent on the extraction steam pressure and the steam pressure at the inlet to the turbine. A typical power-to-heat ratio characteristic is shown in Fig. 9.23. When the extraction steam pressure increases, the power-to-heat ratio reduces for a particular turbine inlet pressure. Therefore, to increase the power-to-heat ratio for a particular extraction steam pressure, the inlet pressure to the turbine has to be increased.

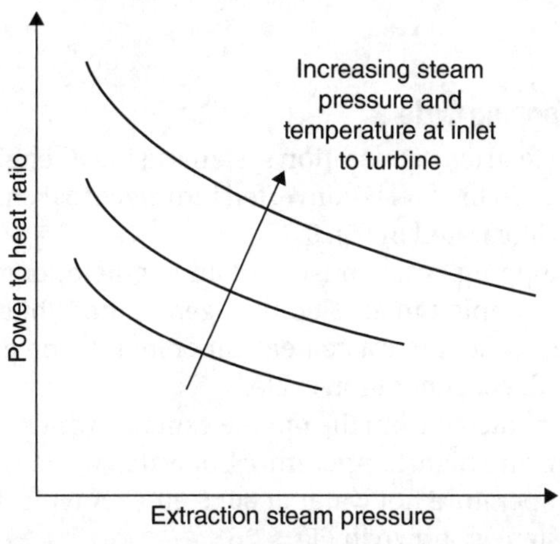

FIGURE 9.23 Power-to-heat ratio characteristic for an extraction-condensing steam turbine.

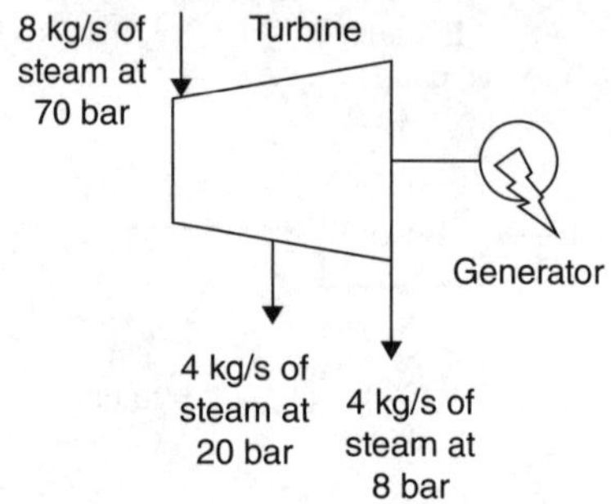

FIGURE 9.24   Diagram for Example 9.10.

Example 9.10   A steam turbine is used to generate electrical energy. Steam at 8 kg/s is supplied at 70 bar and 450°C to the turbine. Then, steam at 4 kg/s is extracted at 20 bar and 250°C, while the remaining steam leaves the turbine at 8 bar and 200°C (Fig. 9.24). Compute the amount of power that can be generated by the turbine if the turbine thermal efficiency is 80%.

Solution

Enthalpy of steam at inlet to turbine = 3287 kJ/kg (h_g at 70 bar and 450°C)

Enthalpy of steam extracted at 20 bar = 2904 kJ/kg (h_g at 20 bar and 250°C)

Enthalpy of steam extracted at 8 bar = 2840 kJ/kg (h_g at 8 bar and 200°C)

Amount of heat remaining for power generation = heat in input steam − heat in extracted steam

$$= 8 \text{ kg/s} \times 3287 \text{ kJ/kg} - [(4 \text{ kg/s} \times 2904 \text{ kJ/kg}) + (4 \text{ kg/s} \times 2840 \text{ kJ/kg})]$$

$$= 3320 \text{ kW}$$

Total amount of power that can be generated by the steam turbine = 3320 kW × 0.8 = 2656 kW   ▲

9.7.4   Engine Topping Cycle

In conventional electricity generation systems using IC engines, only about 30% to 40% of the input energy in the fuel is converted into electrical energy. The remaining energy is lost as heat, as illustrated in Fig. 9.25.

The engine topping cycle consists of an engine operating using fuel such as gas, diesel, or fuel oil coupled to an electrical generator. While electrical energy is generated by this basic cycle, the waste heat generated by the engine is recovered to serve thermal loads in the cogeneration cycle.

Heat can be extracted from the engine exhaust, which is usually in excess of 450°C to produce steam and high-temperature hot water, while jacket cooling can be used to produce low-temperature hot water at 80 to 90°C. A typical arrangement of an engine cogeneration system is shown in Fig. 9.26.

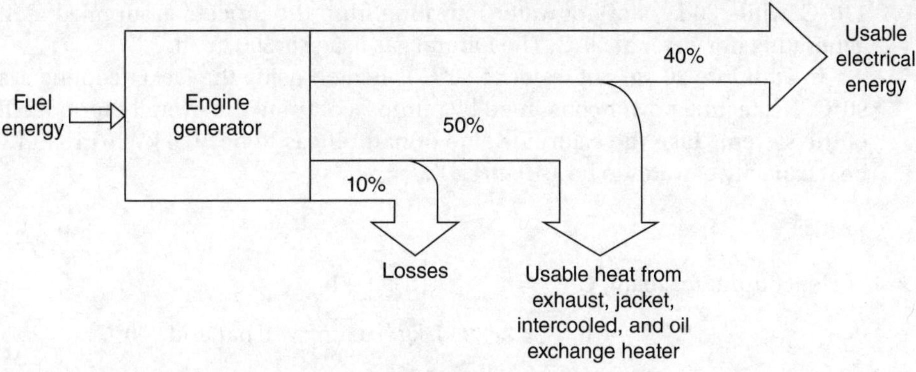

FIGURE 9.25 Heat balance for a typical engine generator.

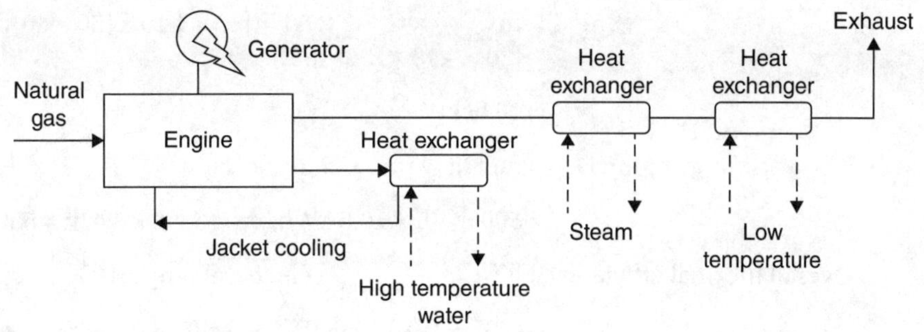

FIGURE 9.26 Typical IC engine cogeneration system.

Example 9.11 As shown in Fig. 9.27, an IC engine operating on natural gas produces a net power output of 2 MW. Heat is recovered from the engine exhaust as well as the jacket cooling system. Steam at 1.5 tons/h at 10 bar and 200°C and hot water at 50°C are produced from the exhaust gas. Feedwater for generating steam is supplied at

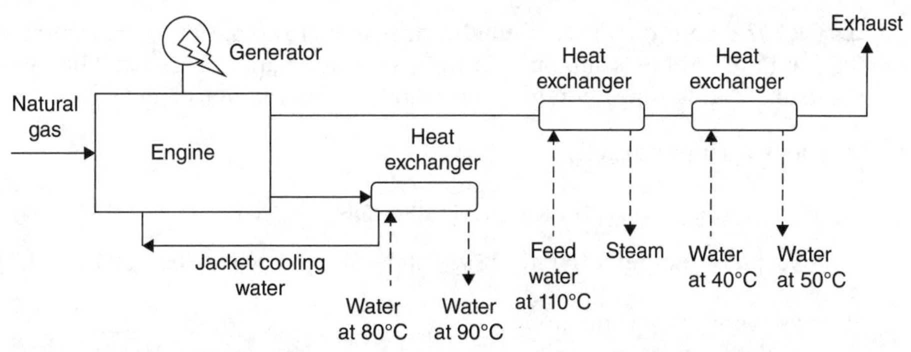

FIGURE 9.27 Diagram for Example 9.11.

110°C, while 20 kg/s of hot water returning from the process is supplied at 40°C for generating hot water at 50°C. The natural gas usage is 500 m³/h.

In addition, 20 kg/s of water at 80°C is heated using the jacket cooling system to 90°C. Neglecting power consumed by pumps, compute the overall thermal efficiency of the system. Take the calorific value of natural gas to be 10.6 kWh/m³ and specific heat capacity of water to be 4.19 kJ/kg·K.

Solution

$$\text{Heat output to steam, } Q_{steam} = m_{steam} \times (h_{steam} - h_{feedwater})$$

$$h_{steam} = 2829 \text{ kJ/kg (steam at 10 bar and 200°C)}$$

$$h_{feedwater} = 4.19 \times 110 = 461 \text{ kJ/kg } (h_f \text{ at } 110°C)$$

$$Q_{steam} = (1500/3600) \text{ kg/s} \times (2829 - 461) = 987 \text{ kW}$$

$$Q_{hot\ water} = m_{hot\ water} \times C_{pwater} \times \Delta T = [20 \times 4.19 \times (50 - 40)] + [20 \times 4.19 \times (90 - 80)]$$

$$Q_{hot\ water} = 1676 \text{ kW}$$

$$\text{Energy input, } Q_{fuel} = \text{fuel flow rate} \times \text{calorific value}$$

$$= 500 \text{ m}^3/\text{h} \times 10.6 \text{ kWh/m}^3 = 5300 \text{ kWh/h} = 5300 \text{ kW}$$

$$\text{Overall thermal efficiency} = [Q_{steam} + Q_{hot\ water} + \text{electrical output}]/Q_{fuel}$$

$$= (987 + 1676 + 2000)/5300 = 0.88 \text{ (or 88\%)} \quad \blacktriangle$$

9.7.5　Combined Cycle Topping

The combine cycle power generation cycle is a combination of the gas turbine and steam turbine power generation cycles, as explained in Sec. 9.5. In combined cycle cogeneration systems, the steam from the steam turbine is supplied for process heating applications rather than condensing in a condenser, as shown in Fig. 9.28. Combined cycle cogeneration systems are used for high-capacity applications and can achieve an overall thermal efficiency of 90% to 95%.

Example 9.12　Consider the combined cycle plant in Example 9.8. If steam after expanding from the turbine is supplied for process heating applications at 2 bar and 150°C (instead of condensing), compute the overall system thermal efficiency.

Solution　Gas turbine cycle:

$$\text{Input power} = 100 \times 1.15 \times (1300 - 589) = 82 \text{ MW}$$

$$\text{Net power output} = 100 \times [1.15 \times (1300 - 839) - 1.005 \times (589 - 305)] = 24.4 \text{ MW}$$

Heat recovery steam generator:

$$\text{Mass flow rate of steam, } m_{steam} = 16.8 \text{ kg/s}$$

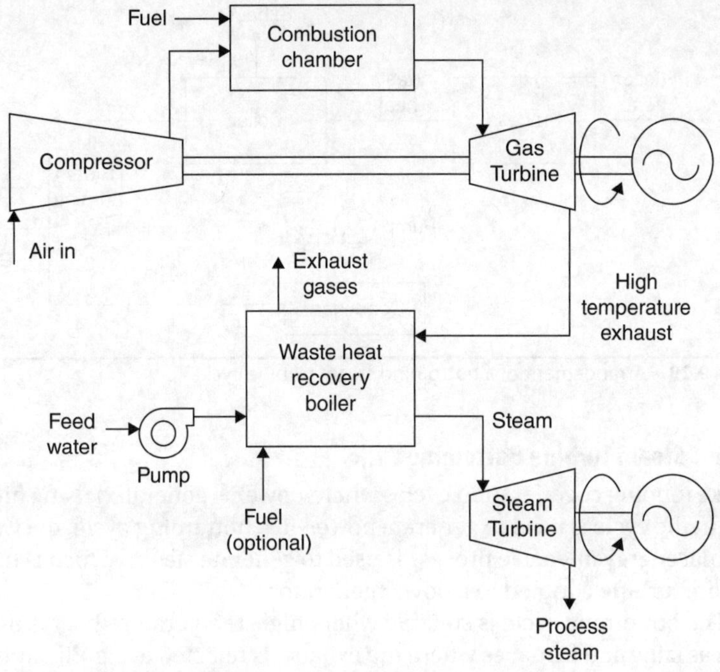

FIGURE 9.28 Arrangement of a combined cycle cogeneration system.

Power output from steam turbine:

$$\text{At inlet to steam turbine, } h_{inlet} = 2846 \text{ kJ/kg (superheated steam at 7 bar and 200°C)}$$

$$\text{At exit of turbine, } h_{outlet} = 2770 \text{ kJ/kg (superheated steam at 2 bar and 150°C)}$$

$$\text{Heat used for process applications} = m_{steam} \times (h_{outlet} - h_f \text{ at 7 bar})$$

$$= 16.8 \times (2770 - 697) = 34{,}826 \text{ kW}$$

$$= 34.8 \text{ MW}$$

$$\text{Enthalpy after pump} = 697 + 0.001 \text{ m}^3/\text{kg} \times (700 \text{ kPa} - 200 \text{ kPa})$$

$$= 697.5 \text{ kJ/kg.}$$

$$\text{Net power output from steam turbine} = \text{output from steam turbine} - \text{power input to pump}$$

$$= 16.8 \times [(2846 - 2770) - (697.5 - 697)]$$

$$= 1.27 \text{ MW}$$

$$\text{Overall plant efficiency} = (\text{output from gas turbine} + \text{output from steam turbine})/\text{gas turbine input power}$$

$$= (24 \text{ MW} + 1.27 \text{ MW} + 34.8 \text{ MW})/82$$

$$= 0.73 \text{ (or 73\%)} \quad \blacktriangle$$

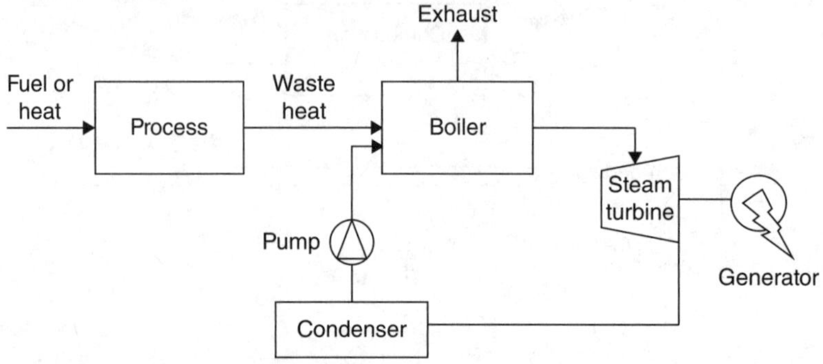

FIGURE 9.29   Arrangement of a bottoming steam turbine cycle.

9.7.6   Steam Turbine Bottoming Cycle

Unlike topping cogeneration cycles where power is generated using fuel and the waste heat in the cycle is used to generate power, in a bottoming cycle, the waste heat from a thermal energy intensive process is used to generate steam, which is then expanded in a steam turbine coupled to a power generator.

The bottoming cycle is suitable where high-temperature heat is used in processes such as kilns and furnaces where the exhaust is rejected at significantly high temperatures. A typical bottoming cycle with a steam turbine is shown in Fig. 9.29.

Such steam turbine bottoming cycles are suitable for petrochemical plants and others, such as cement, steel, and ceramic industries.

9.8   Applications and Considerations

9.8.1   Potential Applications

There are numerous applications for combine heat and power systems. They can generally be categorized into the following three types:

Demand for Both Heat and Electricity

Many industrial facilities, hospitals, and hotels require both electricity and heat energy in large quantities. In such situations, CHP systems can be used to produce both the electrical and thermal energy demands of the facility.

Availability of High-Temperature Waste Heat

Many industrial plants such as cement, glass, ceramic, and chemical manufacturing plants and oil refineries use high-temperature heat for various processes. This waste heat can then be used to generate electricity using a waste heat boiler and a steam turbine.

Locations with No Grid Connectivity

Less-developed countries, remote locations in developed countries, and small resort islands may not have grid connectivity. Therefore, electricity requirements in such locations would normally be provided using stand-alone generation systems. If the

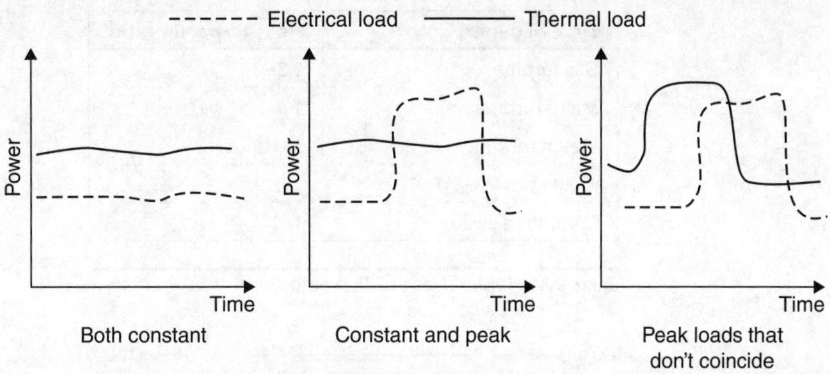

Figure 9.30   Examples of thermal and electrical energy load profiles.

facility also has thermal loads, then instead of rejecting the waste heat from the electricity generation process to the environment, it can be recovered and used to meet part or all of its thermal needs.

9.8.2   Factors Influencing Choice of System

Load Profile

When considering the possibility of using a CHP system, one of the most important factors is the load profile of the thermal and electrical energy loads.

The load profiles may be constant throughout the day or consist of a base load and a peak load, as shown in Fig. 9.30. If both the thermal and electrical energy loads are constant throughout the day, a suitable system can be chosen to match both the loads.

If both the loads have peaks that coincide, then it may be possible to select a system to match both the base and peak load operations.

However, if only one load has a peak or both loads have peaks but do not coincide, then the CHP system will need to be selected based on the base load.

Heat-to-Power Ratio

Another important factor to consider is the ratio of the thermal and electrical energy loads. The heat-to-power ratio is defined as the ratio of heat demand to the electricity demand of a facility.

$$\text{Heat-to-power ratio} = (\text{heat recovered/electricity generated}) = (H/E) \tag{9.9}$$

The operating heat-to-power ratio varies from one CHP system to another, and some typical values are tabulated in Table 9.1.

$$\text{The overall system thermal efficiency, } \eta_{th} = (E + H)/F \tag{9.10}$$

where F = fuel input energy.

The overall thermal efficiency improves with increase in the heat-to-power ratio (Fig. 9.31).

Type of generation	Heat-to-power ratio
Gas turbine	1.5–3
Micro turbine	1.4–2.5
Steam turbine	3–10
IC engine	1–2.0
Combined cycle	1–1.7

TABLE 9.1   Typical Heat-to-Power Ratios

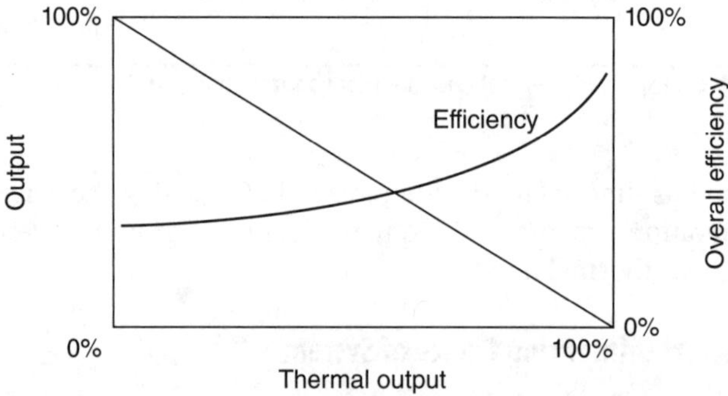

FIGURE 9.31   Relationship between heat-to-power ratio and overall efficiency.

Consider the steam turbine power generation system described in Example 9.4. The overall system thermal efficiency was computed to be 34%. If we take the boiler efficiency as 90%, when 1000 kW of input fuel is supplied, 100 kW will be lost at the boiler while the cycle will generate 340 kW of electricity. The remaining 560 kW of heat energy would be discarded at the condenser (Fig. 9.32).

Now if the system is converted to a CHP system with all the heat recovered after the steam turbine, as shown in Fig. 9.33, the overall efficiency can improve to about 90%.

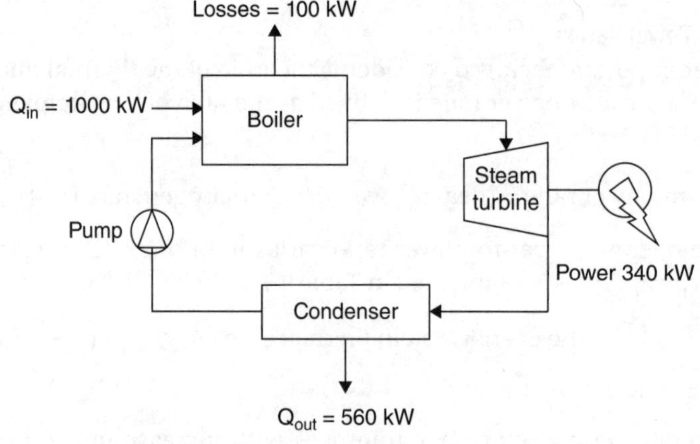

FIGURE 9.32   Power generation without heat recovery.

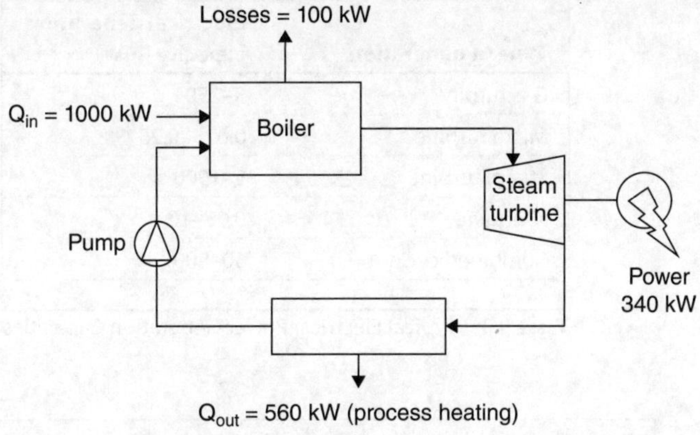

FIGURE 9.33 Power generation with heat recovery.

Amount of heat recovered	H/E ratio	Overall thermal efficiency (E + H)/F
560 kW	1.6	0.9
420 kW	1.2	0.76
280 kW	0.8	0.62
140 kW	0.4	0.48
0 kW	0	0.34

TABLE 9.2 Variation in Efficiency When Different Amounts of Heat is Extracted

The heat-to-power ratio and overall thermal efficiency for different waste heat extraction rates are shown in Table 9.2. As shown in the table, the overall thermal efficiency reduces from the highest value when all the available heat is utilized to the lowest value, which is the efficiency of the pure electricity generation cycle.

Therefore, the higher the heat-to-power ratio, the better the overall thermal efficiency. Since different applications have diverse heat requirements, the most suitable CHP system for a particular application needs to be selected considering multiple factors such as capacity, load profile, fuel availability, and capital cost.

Capacity

Selection of a suitable CHP system for a particular application also depends on the required generation capacity. Table 9.3 provides a summary of the typical size of various electrical power generation systems.

Microturbines are the only option available for small-capacity systems. IC engines and gas turbines are the common choice for medium-sized applications, while combined cycle and steam turbines are used for high-capacity systems.

Type of generation	Electrical generation capacity (MW)
Gas turbine	3–250
Micro turbine	0.025–0.25
Steam turbine	5–1500
IC engine	0.05–10
Combined cycle	50–500

TABLE 9.3 Typical Electrical Power Generation Capacities

Type of generation	Typical fuels
Gas turbine	Natural gas, liquid fuels
Micro turbine	Natural gas
Steam turbine	Solid, liquid, and gaseous fuels (for boiler)
IC engine	Gaseous and liquid fuels
Combined cycle	Gaseous and liquid fuels

TABLE 9.4 Fuels Used by Different Power Generation Systems

Fuel Availability and Cost

The economic feasibility of a CHP system is highly dependent on the availability of cheap fuel. In addition, CHP systems require continuous supply of fuel and, hence, the reliability of the source of fuel also needs to be considered when selecting a system. Table 9.4 summarizes the various types fuels of fuels commonly used to operate the different CHP systems.

9.8.3 System Design Strategies

The two most common strategies used when designing CHP systems are called electricity matching and heat matching.

Electricity matching involves designing the system to match the electrical load of the facility. The heat produced by the system is used to meet the various thermal loads of the facility. If the heat generated is not sufficient to meet the total thermal load, auxiliary boilers would need to be used. Similarly, if the thermal load is lower than the heat generated, part of the heat would need to be wasted by rejecting to the environment.

Heat matching, on the other hand, involves designing the system to match the thermal load of the facility. When using heat matching, the electrical demand should be equal or less than the electricity generation capacity of the system. Electricity from the grid could be used to supplement when the electrical load exceeds the generation capacity. For systems where the electrical generation capacity exceeds the load, depending on local regulations, the excess electricity could be exported to the grid or to neighbouring facilities.

9.9   Tri-Generation Systems

Tri-generation is the combined generation of electricity, heat, and cooling, all simultaneously produced from a single fuel source. It is also often referred to as combined heat power and cooling (CHPC). Tri-generation can be considered as an extension of cogeneration where part of the waste heat is utilized for producing cooling.

Figure 9.34 illustrates the main components of a tri-generation system, which consists of a conventional power generation system together with a waste heat recovery system for steam generation and an absorption chiller for generating chilled water.

The power generation system can be a steam turbine, gas turbine, IC engine, or microturbine. Figure 9.35 shows the arrangement of a typical tri-generation system using an IC engine where hot water from jacket cooling is used for operating an

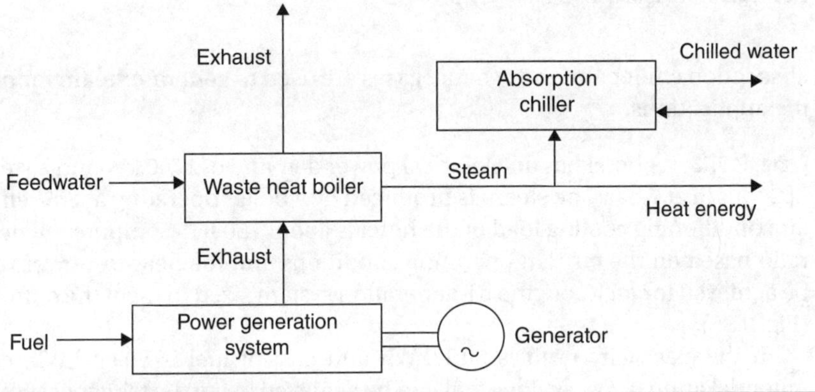

FIGURE 9.34   Arrangement of a tri-generation system.

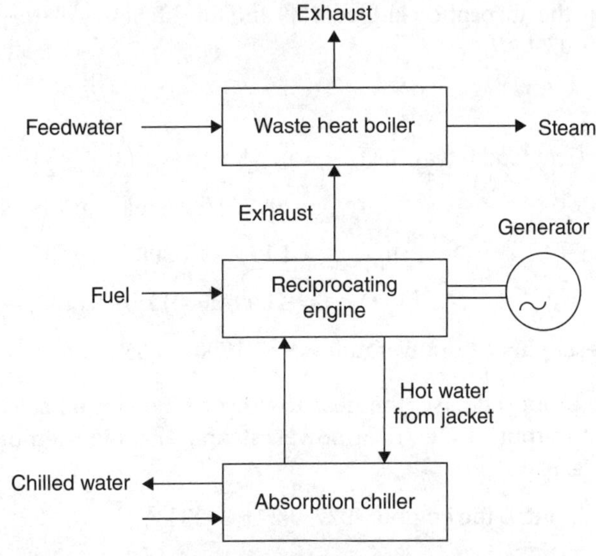

FIGURE 9.35   Arrangement of an IC engine tri-generation system.

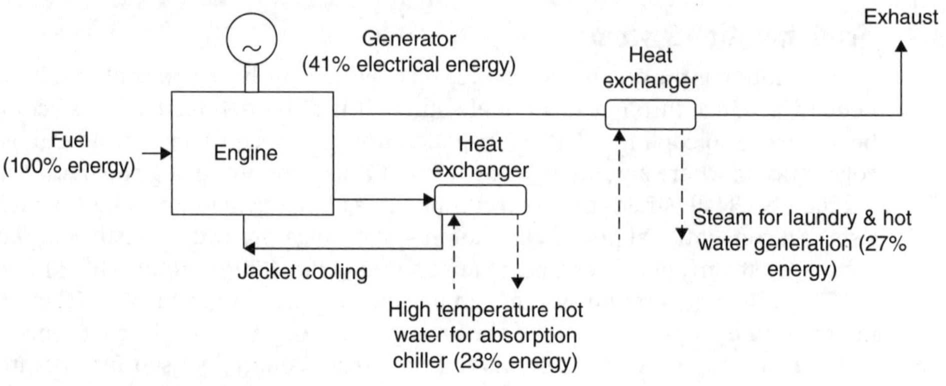

FIGURE 9.36   Diagram for Example 9.13.

absorption chiller and the exhaust gases are used to generate steam for process heating applications.

Example 9.13   A hotel has an electrical power demand of 1700 kW and a steam demand of 2 tons/h at 7 bar. The steam is produced by a boiler operating at 88% efficiency. The air conditioning cooling load of the hotel is about 750 RT. Compute the heat-to-power ratio based on the current operating conditions and the heat-to-power ratio that can be achieved for an IC engine tri-generation system sized to meet the entire steam load (Fig. 9.36).

If the electricity tariff is \$0.1/kWh and cost of fuel is \$0.04/kWh, compute the potential annual cost savings that can be achieved using the tri-generation system.

Feedwater is supplied at 95°C, the specific heat capacity of water is 4.19 kJ/kg·K, and the conversion factor for cooling is 1 RT = 3.517 kW. The COP (cooling kW/heat input kW) of the absorption chiller is 0.9 and the efficiency of the present air conditioning chiller is 0.7 kW/RT.

Solution

$$\text{Heat load (steam)}, Q_{steam} = m_{steam} \times (h_{steam} - h_{feedwater})$$

$$h_{steam} = 2067 \text{ kJ/kg (steam at 7 bar from Table 5.1)}$$

$$h_{feedwater} = 4.19 \times 95 = 398 \text{ kJ/kg } (h_f \text{ at } 95°C)$$

$$Q_{steam} = (2000/3600) \text{ kg/s} \times (2067 - 398) = 927 \text{ kW}$$

$$\text{Present heat-to-power ratio} = 927/1700 = 0.55$$

For the proposed case, the heat-to-power ratio can be calculated using the percentages of output of electrical power, steam, and high-temperature hot water, as shown in Fig. 9.36.

The heat input to the engine = 927/0.27 = 3433 kW

Electrical power output = 3433 × 0.41 = 1407 kW

Heat output as high-temperature hot water $= 3433 \times 0.23 = 790$ kW

Heat-to-power ratio for the tri-generation system $= (927 + 790)/1407 = 1.22$

Amount of cooling that can be produced by the absorption chiller using this high-temperature hot water $= 790 \times \text{COP} = 790 \times 0.9 = 711$ kW $= 711/3.517 = 202$ RT

When cooling is produced by the absorption chiller, the reduction in electrical load (currently drawn by electric chillers) $=$ cooling load $\times$ efficiency $= 202$ RT $\times 0.7$ kW/RT $= 141$ kW

Electrical load of the hotel with the tri-generation system $= 1700 - 141 = 1559$ kW

Electrical power produced by the tri-generation system $= 1407$ kW

Amount of electrical power demand to be obtained from the grid $= 1559 - 1407 = 152$ kW

Present fuel consumption by the boiler $=$ steam output in kW/boiler efficiency $= 927/0.88 = 1053$ kW

Fuel usage by the new generator $= 3433$ kW

Increase in fuel usage $= 3433 - 1053 = 2380$ kW

Extra cost of fuel per day (24 h) $= 2380$ kW $\times 24$ h $\times \$0.04$/kWh $= \$2285$

Present cost of electricity per day $= 1700$ kW $\times 24$ h $\times \$0.1$/kWh $= \$4080$

Grid electricity cost with new system per day $= 152$ kW $\times 24$ h $\times \$0.1$/kWh $= \$365$

Overall savings per day $= \$4080 - \$365 - \$2285 = \$1430$

Annual cost savings $= \$1430 \times 365 = \$521,950$ ▲

CHAPTER 10

Financial Analysis

10.1 Introduction

Other than simple energy-saving measures such as switching off lighting and changing the temperature set-point of air-conditioned spaces, most energy management projects require financial investment to achieve energy savings.

Typical costs associated with energy management projects include the cost of purchase and installation of new equipment, maintenance of the new equipment, and training of staff to operate and maintain them.

Therefore, the decision whether to implement an energy-saving project is often made after financial evaluation of the investment cost and potential benefits that will result from it.

Financial analysis is a useful tool in energy management not only to evaluate the financial viability of projects, but also to compare alternative projects, long-term budgeting, and cash flow planning.

This chapter contains a description of financial evaluation criteria that are commonly used for assessing the feasibility of energy management projects.

10.2 Simple Payback Period

Simple payback period is merely an estimation of how long it will take to recover the initial investment and is defined as the ratio of the initial investment to the annual savings.

$$\text{Simple payback period} = \text{total investment cost/annual savings} \tag{10.1}$$

Example 10.1 The investment cost for new equipment is $500,000 and the resulting cost savings is estimated to be $100,000 a year. Compute the simple payback period for this project.

Solution

$$\text{Simple payback period} = 500,000/100,000 = 5 \text{ years} \quad \blacktriangle$$

The main advantages of this financial evaluation criterion are that it is very simple to use and that it gives an immediate indication of how long it will take to recover the initial investment.

Year	Project A	Project B
Year 0	(500,000)	(570,000)
Year 1	200,000	220,000
Year 2	200,000	220,000
Year 3	200,000	220,000
Year 4	200,000	220,000

TABLE 10.1   Data for Example 10.2

However, the simple payback period evaluation criterion does not consider the time value of money and also does not account for cash flows after the payback period. Therefore, it is useful only when the project time scale is short where the effect of "time value of money" (discussed later) is not significant.

Example 10.2   The respective cash flows for two projects, A and B, are shown in Table 10.1. Compute the simple payback period and total net cash flow resulting from each project.

Solution

$$\text{Simple payback period for Project A} = 500{,}000/200{,}000 = 2.5 \text{ years}$$

$$\text{Simple payback period for Project B} = 570{,}000/220{,}000 = 2.6 \text{ years}$$

$$\text{Total net cash flow for Project A} = (200{,}000 \times 4) - 500{,}000 = \$300{,}000$$

$$\text{Total net cash flow for Project B} = (220{,}000 \times 4) - 570{,}000 = \$310{,}000$$

The simple payback period for Project A is shorter than for Project B, but the total net cash flow for Project B is higher than for Project A.   ▲

The above example helps to illustrate that the simple payback criterion should not be used solely to make investment decisions as projects with longer payback periods may provide higher net gains over the life of a project.

10.3   Return on Investment

Return on investment (ROI) expresses the annual return from the project as a percentage of the investment cost.

$$\text{ROI} = (\text{annual net cash flow/investment cost}) \times 100\% \tag{10.2}$$

For a project to proceed, the ROI should be higher than the cost of capital (interest rate). The higher the difference between ROI and interest rate, the better the project.

This method of financial evaluation is often used to choose between various options available for a particular project. Like the simple payback period method, it is easy to use. It is also able to provide an indication of the return from an investment in percentage, which can be compared with the cost of capital.

The main disadvantages of this evaluation criterion are that it does not consider the time value of money and variable project cash flows.

Example 10.3   New equipment A will cost $500,000 and will provide savings of $80,000 a year. Another new equipment B will cost $650,000 but will provide savings of $85,000 a year. Compare the ROI for the two options.

Solution

$$ROI \text{ for equipment A} = (80{,}000/500{,}000) \times 100 = 16\%$$

$$ROI \text{ for equipment B} = (85{,}000/650{,}000) \times 100 = 13\%$$

The ROI for equipment A is 3% higher than for equipment B.   ▲

10.4   Cash Flow Analysis

The net annual cash flows resulting from a project may not be uniform due to changes in factors such as energy cost, maintenance cost, and utilization hours of the new system.

Consider Table 10.2, which consists of the net annual net cash flows for the first five years for a particular project. These cash flows are also illustrated in Fig. 10.1.

Therefore, for long-term financial planning and evaluation of projects, it is necessary to prepare a cash flow tabulation as illustrated in Example 10.4.

Year	Net annual cash flows	Net cumulative cash flows
Year 0	(500,000)	(500,000)
Year 1	100,000	(400,000)
Year 2	150,000	(250,000)
Year 3	150,000	(100,000)
Year 4	70,000	(30,000)
Year 5	150,000	120,000

TABLE 10.2   Annual Net Cash Flows and Cumulative Cash Flows for a Project

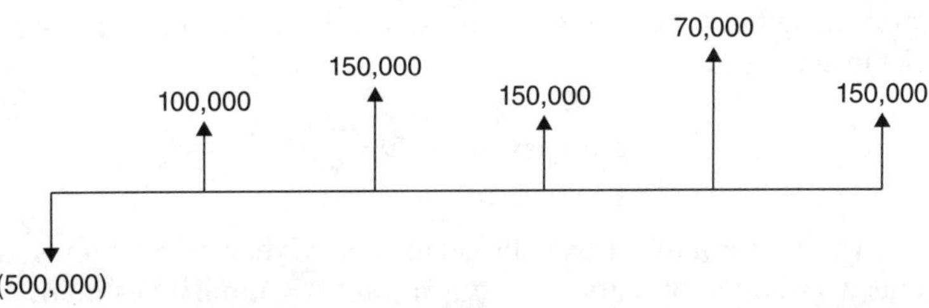

FIGURE 10.1   Illustration of net annual cash flows for the project (Table 10.2).

Example 10.4   The investment cost for new equipment is $250,000. The expected annual energy savings is $50,000 based on the present electricity tariff.

There will be no maintenance cost for the first year of operation of the equipment as it will be covered by a 1-year comprehensive warranty. Thereafter, the annual maintenance cost will be $5000 a year. However, the equipment will need a major overhaul every 5 years and will cost an additional $5,000 in years 5, 10, etc.

Assuming that the electricity tariff will increase by 5% a year from the second year of operation, prepare a cash flow tabulation for the first 10 years of operation for this new equipment.

Solution   The solution is expressed in a tabular format (Table 10.3).   ▲

10.5   Time Value of Money

Cash flows (expenditure and savings) will result during the expected life of the project. The typical project life would be more than 10 years. The monetary value of cash flows resulting during different years will not be the same due to factors such as interest rates and inflation. Hence, when assessing the feasibility of a project, all the present and future cash flows should to be converted to a common value base.

Discounting is used to convert all the future cash flows into the present value (PV) of money. If $1.00 is deposited in a bank account which pays 5% interest, the money deposited will be worth $1.05 in one year. Therefore, the money value of $1.05 one year from now has a present value of $1.00. Similarly, the value of $1.00 in one year's time has a present value of $0.95 (i.e., $1/1.05 = 0.95$).

$$\text{Present value PV} = FV/(1+i)^n \qquad (10.3)$$

where FV = future value of cash flow at the end of year "n", i = interest rate or discounting rate, and n = number of years in the future.

Example 10.5   Calculate the PV of a cash flow of $100,000 that will result in the second year, taking the discounting rate to be 5%.

Solution

$$\text{Present value PV} = 100,000/(1+0.05)^2 = \$90,703 \qquad ▲$$

10.6   Net Present Value

Net present value is the sum of the present values of all cash flows resulting over the entire life of the project.

$$\text{Net present value NPV} = \sum_1^n \frac{Fv_n}{(1+i)^n} - I_0 \qquad (10.4)$$

where FV_n = future value of cash flow at the end of year "n", i = interest rate or discounting rate, n = number of years in the future, and I_0 = initial investment.

If $NPV > 0$, the investment would add value to the company and the project may be accepted.

	Year 0	Year 1	Year 2	Year 3	Year 4	Year 5	Year 6	Year 7	Year 8	Year 9	Year 10
Capital cost	(250,000)										
Energy savings		50,000	52,500	55,125	57,881	60,775	63,814	67,005	70,355	73,873	77,566
Maintenance cost			(5,000)	(5,000)	(5,000)	(10,000)	(5,000)	(5,000)	(5,000)	(5,000)	(10,000)
Net cash flows	(250,000)	50,000	47,500	50,125	52,881	50,775	58,814	62,005	65,355	68,873	67,566
Cumulative cash flows	(250,000)	(200,000)	(152,500)	(102,375)	(49,494)	1,282	60,096	122,100	187,455	256,328	323,895

Table 10.3 Annual Net Cash Flows and Cumulative Cash Flows for Example 10.4

Year	Project A	Project B
Year 0	(500,000)	(700,000)
Year 1	200,000	260,000
Year 2	200,000	260,000
Year 3	200,000	260,000
Year 4	200,000	260,000

TABLE 10.4   Annual Net Cash Flows Data for Example 10.6

If NPV < 0, the investment would diminish value from the company and the project would normally be rejected.

If NPV = 0, the investment would not gain or lose value for the company and the decision whether to proceed or reject the project would depend on other factors.

Example 10.6   Calculate the NPV for the following two projects (Table 10.4) taking the discounting rate to be 5%.

Solution

Project A:

$$NPV = 200{,}000/(1 + 0.05)^1 + 200{,}000/(1 + 0.05)^2 + 200{,}000/(1 + 0.05)^3$$
$$+ 200{,}000/(1 + 0.05)^4 - 500{,}000$$
$$= 190{,}476 + 181{,}406 + 172{,}768 + 164{,}540 - 500{,}000$$
$$= \$209{,}190$$

Project B:

$$NPV = 260{,}000/(1 + 0.05)^1 + 260{,}000/(1 + 0.05)^2 + 260{,}000/(1 + 0.05)^3$$
$$+ 260{,}000/(1 + 0.05)^4 - 700{,}000$$
$$= 247{,}619 + 235{,}828 + 224{,}598 + 213{,}903 - 700{,}000$$
$$= \$221{,}948 \quad \blacktriangle$$

To simplify the computation of NPV, values of the factor $(1 + i)^n$ are normally provided in reference tables such as Table 10.5.

Example 10.7   Using Table 10.5, compute the PV of a cash flow of \$150,000 that will result in year 10, taking the discounting rate to be 8%.

Solution   From Table 10.5, the multiplication factor for Year 10 and discounting rate of 8% is 0.463.

$$\text{Therefore, the PV} = 150{,}000 \times 0.463 = \$69{,}450 \quad \blacktriangle$$

The NPV can also be easily calculated using the standard function available in spreadsheet software. In Microsoft Excel, the standard function for NPV is

$$NPV \text{ (discounting rate, value range)} + cost$$

Year	Discounting rate								
	1%	3%	5%	6%	8%	10%	12%	15%	20%
1	0.990	0.971	0.952	0.943	0.926	0.909	0.893	0.870	0.833
2	0.980	0.943	0.907	0.890	0.857	0.826	0.797	0.756	0.694
3	0.971	0.915	0.864	0.840	0.794	0.751	0.712	0.658	0.579
4	0.916	0.888	0.823	0.763	0.735	0.683	0.636	0.572	0.482
5	0.951	0.863	0.784	0.747	0.681	0.621	0.567	0.497	0.402
6	0.942	0.837	0.746	0.705	0.630	0.564	0.507	0.432	0.335
7	0.933	0.813	0.711	0.665	0.583	0.513	0.452	0.376	0.279
8	0.923	0.789	0.677	0.627	0.540	0.467	0.404	0.327	0.233
9	0.914	0.766	0.645	0.592	0.500	0.424	0.361	0.284	0.194
10	0.905	0.744	0.614	0.558	0.463	0.386	0.322	0.247	0.162
11	0.896	0.722	0.585	0.527	0.429	0.350	0.287	0.215	0.135
12	0.887	0.701	0.557	0.497	0.397	0.319	0.257	0.187	0.112
13	0.879	0.681	0.530	0.469	0.368	0.290	0.229	0.163	0.093
14	0.870	0.661	0.505	0.442	0.340	0.263	0.205	0.141	0.078
15	0.861	0.642	0.481	0.417	0.315	0.239	0.183	0.123	0.065

TABLE 10.5   Present Value Factors

The data for Project A in Example 10.6 can be arranged in MS Excel as follows:

Row 1	Column B	Column C	Column D	Column E	Column F
Row 2	Year 0	Year 1	Year 2	Year 3	Year 4
Row 3	(500,000)	200,000	200,000	200,000	200,000
Row 4					

The function to compute NPV would be "NPV (0.05,C3:F3) + B3".

The advantages of using the NPV criterion to evaluate projects are that it considers the time value of money, takes into account the variable nature of annual cash flows, and can indicate whether the investment will add value to the company.

Disadvantages of the NPV method are that it needs an estimate of the cost of capital (discounting rate) and provides only the value added in absolute terms (not a percentage) to compare with the overall financial performance of a company.

10.7   Discounting Rate

The discounting rate used should reflect the cost of capital (borrowing cost) and the risk associated with the project, if the investment is to be financed by borrowing.

If the project is to be financed using company funds, the discounting rate should also take into account the expected ROI (opportunity cost) for the company.

A few examples of discounting rates for projects (without considering risk) are listed below:

1. If a project requires funds to be borrowed at 7% interest rate, the discounting rate used should be 7% or higher.

2. If a project is to be financed using internal funds, and alternative investment can bring the company returns of 10%, then the discounting rate used should not be lower than 10%.

3. If a project requires $4 million for which $1 million is to be borrowed at 7% while the remaining $3 million will be paid from the company's own funds, which can yield 10% in an alternative investment, then the weighted average discounting rate $(0.25 \times 7\%) + (0.75 \times 10\%) = 9.25\%$ should be used as the minimum criterion.

10.8   Discounted Payback Period

The discounted payback period method uses the discounted cash flows to compute the payback period for a project and therefore is able to account for time value of money.

Discounted payback period = year before recovery
     + (unrecovered cost at the beginning of the year/cash flow during the year)

Example 10.8 illustrates the computation of the discounted payback period.

Example 10.8   Calculate the discounted payback period for the following project (see Table 10.6), taking the discounting rate to be 7%.

Solution   The solution is expressed in a tabular format below (Table 10.7).

Discounted payback period = year before recovery
     + (unrecovered cost at the beginning of the year/cash flow during the year)   (10.5)

Year	Cash flows
Year 0	(1,500,000)
Year 1	400,000
Year 2	450,000
Year 3	500,000
Year 4	600,000

TABLE 10.6   Data for Example 10.8

Year	Cash flows	Discounted cash flows	Cumulative discounted cash flows
Year 0	(1,500,000)		
Year 1	400,000	373,832	373,832
Year 2	450,000	393,047	766,879
Year 3	500,000	408,149	1,175,028
Year 4	600,000	457,737	1,632,765

TABLE 10.7 Cumulative Discounted Cash Flows for Example 10.8

Since the total investment cost is $1,500,000, the year when this total sum is recovered is year 4 (cumulative discounted cash flow exceeds $1,500,000). Therefore, the year before recovery is year 3.

$$\text{Discounted payback period} = 3 + (1,500,000 - 1,175,028)/457,737$$
$$= 3 + 0.71$$
$$= 3.71 \text{ years} \quad \blacktriangle$$

10.9 Internal Rate of Return

Internal rate of return (IRR) calculates the rate of return an investment is expected to make. The IRR is the interest (discounting) rate at which the NPV becomes zero (i.e., discounted benefits are equal to the discounted costs).

$$\text{NPV} = 0 = FV_0/(1+i)^0 + FV_1/(1+i)^1 + FV_2/(1+i)^2 + \cdots + FV_n/(1+i)^n$$

where FV_n = future value of cash flow at the end of year "n" (positive for savings and negative for expenditure), i = interest rate or discounting rate, and n = number of years in the future.

The value of "i" (discounting rate) needs to be computed by trial and error. For a project to be feasible, the IRR should be higher than the "hurdle rate" (cost of capital) for the company. The higher the IRR, the better the expected return from the project. Similarly, when evaluating different investment options, the one with the highest rate of return would normally be chosen.

Example 10.9   For the following project cash flow (Table 10.8), compute the IRR.

Solution

$$\text{NPV} = 0 = 200,000/(1+i)^1 + 200,000/(1+i)^2 + 200,000/(1+i)^3$$
$$+ 200,000/(1+i)^4 - 500,000$$

To solve by trial and error, initially use i = 10%.

$$\text{NPV} = 200,000/(1+0.1) + 200,000/(1+0.1)^2 + 200,000/(1+0.1)^3$$
$$+ 200,000/(1+0.1)^4 - 500,000$$
$$= 181,818 + 165,289 + 150,263 + 136,603 - 500,000$$
$$= \$133,973$$

Since the NPV is positive, it is necessary to use a higher value of i.

Year	Cash flow
Year 0	(500,000)
Year 1	200,000
Year 2	200,000
Year 3	200,000
Year 4	200,000

TABLE 10.8   Data for Example 10.9

Therefore, try, i = 20%.

$$NPV = 200,000/(1 + 0.2) + 200,000/(1 + 0.2)^2 + 200,000/(1 + 0.2)^3$$
$$+ 200,000/(1 + 0.2)^4 - 500,000$$
$$= 166,667 + 138,889 + 115,741 + 96,451 - 500,000$$
$$= \$17,748 \text{ (still positive)}$$

Still the NPV is positive so next try i = 25%.

$$NPV = 200,000/(1 + 0.25) + 200,000/(1 + 0.25)^2 + 200,000/(1 + 0.25)^3$$
$$+ 200,000/(1 + 0.25)^4 - 500,000$$
$$= 160,000 + 128,000 + 102,400 + 81,920 - 500,000$$
$$= \$-27,680$$

Since the NPV becomes negative when i = 25%, the IRR is between 20% and 25%. Further computations using discounting rates of between 20% and 25% will yield the actual IRR to be 22%. ▲

Although calculating the IRR by trial and error can be tedious, it can be computed relatively easily using the standard IRR function available in spreadsheet software.

In MS Excel, the standard function is "IRR (value range, guess)" where the "value range" refers to the annual net cash flows, while the "guess" refers to the starting guess for the iterative calculation. If the guess value is left blank, then a default value of 10% is used.

If the cash flows for Example 10.9 are arranged as shown below, the standard function to compute the IRR should be "IRR(B3:F3)" (taking the default first guess value).

Row 1	Column B	Column C	Column D	Column E	Column F
Row 2	Year 0	Year 1	Year 2	Year 3	Year 4
Row 3	(500,000)	200,000	200,000	200,000	200,000
Row 4					

Some of the main advantages of using the IRR criterion for financial evaluation of projects are that it takes into account the time value of money, it is able to consider the variable nature of annual cash flows, and is able to provide a rate of return value that can be compared with the company "hurdle rate" (cost of capital) to assist in the decision-making.

Some of the disadvantages of this method are that it requires an estimate of the cost of capital, it does not provide the absolute improvement in company value (provides only a percentage), and cannot allow more than one change in sign in the annual net cash flows.

Since each of the financial assessment criteria described earlier has its own advantages and disadvantages, when evaluating a project, the most appropriate criterion should be used. Often, it may also be necessary to use more than one financial assessment criterion before making a decision.

10.10 Sensitivity Analysis

In financial evaluation of projects, cash flow projections are made using various assumptions for factors such as energy cost, maintenance cost, project investment cost, interest (discounting) rates, and equipment utilization rates.

Since the future values of these variables are uncertain, it is possible that one or more of them can change during the life span of the project, affecting the financial performance of the investment.

Therefore, a sensitivity analysis should be done to understand the possible risk involved by assessing the project financial performance when key factors are varied. For instance, the projected IRR for a particular investment is estimated to be 10% when the hurdle rate for the company is 8%, slight changes to one or more variables can result in the IRR falling below the hurdle rate, which will make the project unviable.

Hence, financial performance criteria such as NPV and IRR should be computed using different values of variables that are likely to change to assess the risk involved.

Examples of factors that may need to be varied in a sensitivity analysis for energy management projects are reductions to electricity tariff (energy cost), increase in capital cost, reduction in the utilization of the proposed new system, and reduction in service life of the new equipment.

A sensitivity analysis using data from Example 10.9 is illustrated below for the following scenarios:

- Case A—5% reduction in annual net cash flows (reduction in energy cost)
- Case B—20% increase in capital cost
- Case C—Reduction in equipment service life by 1 year
- Case D—A combination of 5% reduction in annual net cash flows and a 10% increase in capital cost

10.10.1 Case A—5% Reduction in Annual Net Cash Flows

Row 1	Column B	Column C	Column D	Column E	Column F
Row 2	Year 0	Year 1	Year 2	Year 3	Year 4
Row 3	(500,000)	190,000	190,000	190,000	190,000
Row 4					

Using the standard Excel function "IRR(B3:F3)", the value of IRR = 19%.

10.10.2 Case B—20% Increase in Capital Cost

Row 1	Column B	Column C	Column D	Column E	Column F
Row 2	Year 0	Year 1	Year 2	Year 3	Year 4
Row 3	(600,000)	200,000	200,000	200,000	200,000
Row 4					

Using the standard Excel function "IRR(B3:F3)", the value of IRR = 13%.

10.10.3 Case C—Reduction in Equipment Service Life by 1 Year

Row 1	Column B	Column C	Column D	Column E	Column F
Row 2	Year 0	Year 1	Year 2	Year 3	Year 4
Row 3	(500,000)	200,000	200,000	200,000	—
Row 4					

Using the standard Excel function "IRR(B3:E3)", the value of IRR = 10%.

10.10.4 Case D—Combination of 5% Reduction in Annual Net Cash Flows and 10% Increase in Capital Cost

Row 1	Column B	Column C	Column D	Column E	Column F
Row 2	Year 0	Year 1	Year 2	Year 3	Year 4
Row 3	(500,000)	190,000	190,000	190,000	190,000
Row 4					

Using the standard Excel function "IRR(B3:F3)", the value of IRR = 14%.

The lowest resulting IRR from the above illustrated sensitivity analysis is 10%. If this value of IRR is higher than the hurdle rate for the company and if all the possible risk factors have been considered in the above sensitivity analysis, then this investment can be considered to be acceptable.

10.11 Life Cycle Cost

Most energy management projects involve significant capital expenditure that needs to be recovered over the expected life of the project. Therefore, a complete financial evaluation of a project should account for all key costs and revenues associated with that project.

Typical costs that need to be included in such a life cycle cost (LCC) analysis are the equipment purchase cost, energy cost, water cost (where relevant), maintenance cost, replacement cost, and residual value (if any) of the equipment.

LCC can be expressed as

$$LCC = \text{first cost} + \text{energy cost} + \text{water cost} + \text{maintenance cost} + \text{replacement cost} - \text{residual value}$$

$$(10.6)$$

Since the various cash flows will result at different times in the project life cycle, all values should be converted into the PV before adding them to compute the LCC.

When comparing alternative projects, the option with the lowest LCC should be chosen for implementation.

For a typical chiller replacement project, costs that need to be included in an LCC analysis are as follows:

- Equipment purchase and installation cost
- Energy cost
- Water cost (if it is a water cooled chiller)
- Maintenance cost
- Refrigerant cost
- Pumping cost (energy cost for the associated chilled water and condenser water pumps as the evaporator and condenser pressure losses can be vastly different for individual chillers)

For a typical lighting replacement project, costs that need to be included in a LCC analysis are as follows:

- Cost of new lamps
- Installation cost of lamps
- Energy cost
- Maintenance cost
- Replacement cost (manpower, scaffolding, etc.)

Similarly, the LCC can be computed for any project by identifying all the costs associated with that particular project, converting all the values into the PV, and summing them up to identify the option which has the lowest total LCC.

Example 10.10   In a water pumping system, the flow rate and pump pressure head are measured to be 900 CMH and 42 m (of water), respectively. The operating pump efficiency and motor efficiency are estimated to be 70% and 88%, respectively.

An investigation to study the possibility of reducing the energy consumption of the pumping system has revealed that the flow rate can be lowered by about 10%.

The following three options are to be considered:

- Install a variable-frequency drive to reduce the pump operating speed. It can be assumed that the new system operating pressure at the lower flow rate can be estimated using the affinity laws (Chap. 3) and that the pump efficiency and motor efficiency remain the same when the pump capacity is reduced. The cost of supply and installation of the variable-frequency drive is estimated to be $4000.

- Reduce the pump impeller diameter to enable the pump to operate at a lower capacity. However, from the pump performance curves, it is expected that the pump efficiency will reduce to 60% but the motor efficiency will remain the same. The cost of reducing the impeller size is $1500.

- Install a new pump suitably sized to meet the new operating point. The cost of the new pump will be $8000. The efficiency of the new pump and motor are expected to be 80% and 94%, respectively.

The pumping system operates 24 hours per day and 365 days a year. The electricity tariff is $0.10/kWh. The maintenance cost for the existing pump is $6000 a year. The maintenance cost for the new pump (option 3) will remain the same, except that there will be no maintenance cost in the first year as it will be provided free of charge under the equipment warranty. In option 2, there will be an additional maintenance cost of $3000 a year for the variable-frequency drive.

Perform an LCC analysis for the above three options, taking into consideration only the cost of implementing the improvement measure, operating energy cost, and the maintenance cost (taking the residual value to be zero). The electricity tariff is expected to rise at 2% a year.

Solution   Using Eq. (3.2), the pump motor energy consumption can be estimated as follows:

Option 1 (variable-frequency drive):

$$\text{Flow rate } Q = 900 \times 0.9 \text{ CMH} = 0.225 \text{ m}^3/\text{s}$$

$$\text{Pump head, } \Delta P = 42 \times 0.9^2 \text{ m of water} = 1000 \text{ kg/m}^3 \times 9.81 \text{ m}^2/\text{s} \times 34 = 333{,}736 \text{ N/m}^2$$

$$\text{Pump efficiency} = 70\%$$

$$\text{Motor efficiency} = 88\%$$

Estimated pump power = $(0.225 \times 333{,}736)/0.7 \times 0.88 = 121{,}900$ W = 122 kW

Annual energy consumption = $122 \times 24 \times 365 = 1{,}068{,}720$ kWh/year

Annual energy cost = $1{,}068{,}720 \times 0.1 = \$106{,}872$

Option 2 (changing impeller size):

$$\text{Flow rate } Q = 900 \times 0.9 \text{ CMH} = 0.225 \text{ m}^3/\text{s}$$

$$\text{Pump head, } \Delta P = 42 \times 0.9^2 \text{ m of water} = 1000 \text{ kg/m}^3 \times 9.81 \text{ m}^2/\text{s} \times 34 = 333{,}736 \text{ N/m}^2$$

$$\text{Pump efficiency} = 60\%$$

$$\text{Motor efficiency} = 88\%$$

Estimated pump power = $(0.225 \times 333{,}736)/0.6 \times 0.88 = 142{,}217$ W = 142 kW

Annual energy consumption = $142 \times 24 \times 365 = 1{,}243{,}920$ kWh/year

Annual energy cost = $1{,}243{,}920 \times 0.1 = \$124{,}392$

Option 3 (new pump):

$$\text{Flow rate } Q = 900 \times 0.9 \text{ CMH} = 0.225 \text{ m}^3/\text{s}$$

$$\text{Pump head, } \Delta P = 42 \times 0.9^2 \text{ m of water} = 1000 \text{ kg/m}^3 \times 9.81 \text{ m}^2/\text{s} \times 34 = 333{,}736 \text{ N/m}^2$$

$$\text{Pump efficiency} = 80\%$$

$$\text{Motor efficiency} = 94\%$$

$$\text{Estimated pump power} = (0.225 \times 333{,}736)/0.8 \times 0.94 = 99{,}855 \text{ W} = 99.9 \text{ kW}$$

$$\text{Annual energy consumption} = 99.9 \times 24 \times 365 = 875{,}124 \text{ kWh/year}$$

$$\text{Annual energy cost} = 875{,}124 \times 0.1 = \$87{,}512$$

The LCC for each option can be summarized as follows (for simplicity, time value of money is not considered):

Option 1

	Year 0	Year 1	Year 2	Year 3	Year 4	Year 5	Year 6	Year 7	Year 8	Year 9	Year 10
Implementation cost	4000										
Operating energy cost		106,872	109,009	111,190	113,413	115,682	117,995	120,355	122,762	125,218	127,722
Maintenance cost		9000	9000	9000	9000	9000	9000	9000	9000	9000	9000
Total cost	4000	115,872	118,009	120,190	122,413	124,682	126,995	129,355	131,762	134,218	136,722
Cumulative total cost	4000	119,872	237,881	358,071	480,484	605,166	732,162	861,517	993,279	1,127,497	1,264,219

Option 2

	Year 0	Year 1	Year 2	Year 3	Year 4	Year 5	Year 6	Year 7	Year 8	Year 9	Year 10
Implementation cost	1500										
Operating energy cost		124,392	126,880	129,417	132,006	134,646	137,339	140,086	142,887	145,745	148,660
Maintenance cost		6000	6000	6000	6000	6000	6000	6000	6000	6000	6000
Total cost	1500	130,392	132,880	135,417	138,006	140,646	143,339	146,086	148,887	151,745	154,660
Cumulative total cost	1500	131,892	264,772	400,189	538,195	678,841	822,180	968,265	1,117,153	1,268,898	1,423,558

Option 3

	Year 0	Year 1	Year 2	Year 3	Year 4	Year 5	Year 6	Year 7	Year 8	Year 9	Year 10
Implementation cost	8000										
Operating energy cost		87,512	89,262	91,047	92,868	94,726	96,620	98,553	100,524	102,534	104,585
Maintenance cost		—	6000	6000	6000	6000	6000	6000	6000	6000	6000
Total cost	8000	87,512	95,262	97,047	98,868	100,726	102,620	104,553	106,524	108,534	110,585
Cumulative total cost	8000	95,512	190,774	287,822	386,690	487,416	590,036	694,589	801,113	909,647	1,020,232

Based on the above comparison, option 3 has the lowest LCC and can be considered to be the best option.   ▲

Bibliography

ASHRAE Handbook: Fundamentals. American Society of Heating, Refrigerating and Air-Conditioning Engineers, Inc., Atlanta, GA, 2013.

ASHRAE Handbook: HVAC Systems and Equipment. American Society of Heating, Refrigerating and Air-Conditioning Engineers, Inc., Atlanta, GA, 2012.

ASHRAE Handbook of Refrigeration. American Society of Heating, Refrigerating, and Air-Conditioning Engineers, Inc., Atlanta, GA, 2015.

Bergman, Lavine, Incropera, and DeWitt. *Fundamentals of Heat and Mass Transfer*, 7th ed. John Wiley & Sons, New York, 2011.

Cengel, Yunus A., and Boles, Michael A. *Thermodynamics: An Engineering Approach.* McGraw-Hill, New York, 2014.

Council of Industrial Boiler Owners. *CIBO Energy Efficiency Handbook.* Council of Industrial Boiler Owners, 1997.

Custodio, Jim. The Impact of Rewinding on Motor Efficiency. *Pumps & Systems.* Cahaba Media Group, 2007.

Energy Management of Motor Driven Systems. U.S. Department of Energy, Office of Industrial Technologies, Washington, DC, 2000.

Facts Worth Knowing about Frequency Converters. Danfoss, Nordborg, 1998.

IEC 60034-30 Standard on Efficiency Classes for Low Voltage AC Motors. International Electrotechnical Commission, London, 2008.

Improving Compressed Air System Performance: A Source Book for Industry. Compressed Air Challenge and U.S. Department of Energy, Washington, DC, 2003.

Improving Fan System Performance: A Source Book for Industry. U.S. Department of Energy, Air Movement and Control Association International Inc., Washington, DC, 1989.

Improving Motor and Drive System Performance: A Source Book for Industry. U.S. Department of Energy, Industrial Technologies Program, Washington, DC, 2008.

Improving Pumping System Performance: A Source Book for Industry. U.S. Department of Energy, Office of Energy Efficiency and Renewable Energy, Industrial Technologies Program, Washington, DC, 2006.

Improving Steam System Performance: A Source Book for Industry. U.S. Department of Energy, Industrial Technologies Program, Washington, DC, 2012.

Industrial Refrigeration Best Practices Guide. Northwest Energy Efficiency Alliance, Washington, 2007.

ISO 50001:2011: Energy Management Systems. International Organization for Standardization, Geneva, 2011.

Jayamaha, Lal. *Energy-Efficient Building Systems—Green Strategies for Operation and Maintenance.* McGraw-Hill, New York, 2007.

Krarti, M. *Energy Audit of Building Systems—An Engineering Approach.* CRC Press, Boca Raton, FL, 2000.

NEMA (National Electrical Manufacturers Association). *Standards Publication MG1, 2006.*

Petruzella, Frank D. *Electrical Motors and Control Systems.* McGraw-Hill, New York, 2010.

Pump Selection Software. ITT Belle & Gossette, New York.

Rishel, James B. *HVAC Pump Handbook,* 2nd ed. McGraw-Hill, New York, 1996.

Robertson, C. R. *Fundamental Electrical & Electronic Principles.* Newnes, Oxford, 1993.

Singapore Standard SS530:2006: Code of Practice for Energy Efficiency Standard for Building Services and Equipment. SPRING, Singapore, 2006.

Stoecker, Wilbert F. *Industrial Refrigeration Handbook.* McGraw-Hill, New York, 1998.

The Steam and Condenser Loop. Spirax Sarco Limited, England, 2008.

Training Manual for Motor Driven Systems Module of the Singapore Certified Energy Manager Course (unpublished).

Index